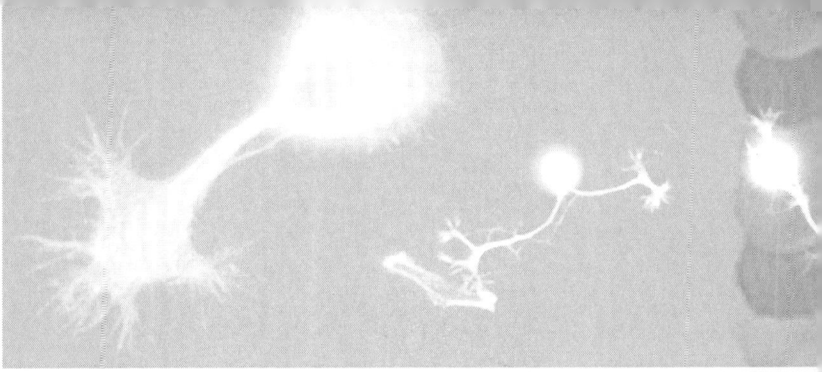

Molecular and Cellular Biomechanics

Molecular and Cellular Biomechanics

edited by

Bradley E. Layton

Pan Stanford Publishing

Published by

Pan Stanford Publishing Pte. Ltd.
Penthouse Level, Suntec Tower 3
8 Temasek Boulevard
Singapore 038988

Email: editorial@panstanford.com
Web: www.panstanford.com

British Library Cataloguing-in-Publication Data
A catalogue record for this book is available from the British Library.

Molecular and Cellular Biomechanics

ISBN 978-981-4316-83-5 (Hardcover)
ISBN 978-981-4613-44-6 (eBook)

Printed in the USA

This book is dedicated to the memory of Prof. Alan Hunt, who taught a wonderful graduate course titled "Cell Mechanics" in the Biomedical Engineering Department at the University of Michigan. Alan's intellect, friendship, and open-mindedness are truly missed.

Contents

Acknowledgments

The authors would like to gratefully acknowledge support from the following sources:

- National Science Foundation CMII 07030000 "Multiscale Structure-Function Relationships of Collagen in the Marine Cyanobacterium Trichodesmium erythraeum"
- National Science Foundation IOS 0950374 "LiT: RUI: Mitigation of Dehydration-Induced Nanomechanical Failure in Arabidopsis thaliana"
- USDA 2008-35100-04413 "Does nanoscale cellulose fibril rearrangement in mesophyll and vascular tissues affect survival rates during dehydration in Eragrostis?"
- Keck Foundation "The Keck Institute for Attofluidic Nanotube-Based Probes"
- National Science Foundation: Kate Rodowicz (Allen) NSF Research Fellow

Preface

This book has been written by engineers and physicists working in various fields of biomechanics. Its intended audience includes upper-level undergraduate students, graduate students, or those generally interested in understanding cellular and molecular mechanics on a more fundamental level. It begins with a general introduction to the scales and terms used in the field of cellular and molecular biomechanics, followed by six chapters, each of which focus on various tissues or cellular systems. Each chapter has a few problems or questions to help the reader dig deeper into the material.

Chapter 1

Introduction

Bradley Layton[a,b] and Kayla Nagle[b]

[a]*Applied Computing and Electronics Department and*
[b]*Biochemistry and Biophysics Program, The University of Montana,*
32 Campus Drive, Missoula, MT 59812, USA

bradley.layton@umontana.edu

1.1 Brief Overview of Numbers and Scales

1.1.1 Number of Cells on the Planet and in Organisms

Just as the atom is the structure that defines the boundary between chemical and nuclear phenomena, the cell serves as the structure that defines the boundary between living and non-living. While it would be difficult to argue that any single molecule within a cell is alive, pluck a single cell at random from an organism, and chances are good that all of the information and machinery required for life will be found there.

Since the vast majority of species on the planet are represented by single-celled organisms, the typical number of cells per organism turns out to be exactly one. This is indeed the case for the estimated 5×10^{30} prokaryotes on the planet (Oren, 2004). But how many cells are there? The total mass of living matter is approximately

Molecular and Cellular Biomechanics
Edited by Bradley E. Layton

ISBN 978-981-4316-83-5 (Hardcover), 978-981-4613-44-6 (eBook)
www.panstanford.com

2000 billion dry tons (Boyle, 2004), or approximately 10,000 billion tons total. If we assume that 50% of this mass is cellular, and that the average cell linear dimension of the average cell is approximately 1 μm, then there are about 10^{31} cells alive right now on the planet, which are represented by approximately 10 million species (Mora et al., 2011) all of which are related via a common ancestor (Dawkins, 2004). Despite their numbers, prokaryotes have been found to be less genetically diverse than insects (Finlay, 2002; Nee, 2003), so genetic diversity is only weakly correlated to the number of cells (Torsvik et al., 2002).

There are some cell-like entities that defy classification. Without a nucleus, and therefore no self-contained means of reproducing, viruses fall into this category (Fuhrman et al., 1993; Breitbart et al., 2007). Red blood cells, having no nucleus of their own, also sit at the classically defined border between living and non-living. Nevertheless, any structure that maintains a boundary with the environment and extracts energy from it to maintain its internal order falls within some broad category of life (Schrödinger, 2012), and can loosely be defined as an "entropy partitioning system": more on this topic when we pay Maxwell's Demon a visit in Chapter 2.

And regardless of their size or genetic complexity, every cell's purpose is to serve its host organism in some capacity. But what are the practical limits for a single organism's size? It was in fact none other than Galileo Galilei, who suggested that there might be a limit to how large organisms could become (Galilei, 1991). Huxley also weighed in on "Being the Right Size" (Huxley, 1993). Stephen J. Gould's so-called "left-wall hypothesis" (Aarssen et al., 2006) also predicts that the distribution of organism masses should fall into a predictable histogram, with the less numerous larger chemotrophic organisms (and species) at the top of the food chain, extracting the hard-won energy from the more numerous prototrophs at the bottom.

During a recent discussion with a neuroscientist colleague, he remarked on what giants humans truly are compared to most organisms. We may not be whales, but it is easy to deduce that there are as many cells in a human heart as there are people in China. There are more neurons in a human head than there are people on the planet, and the number of microorganisms that occupy the various invaginations of the intestines, respiratory system, and various other nooks and crannies of the human body approaches the US national debt in dollars. In this sense, with our hundreds of trillions of cells, each of us resembles a highly organized bipedal mobile civilization.

1.1.2 Number of Cell Phenotypes in an Organism

It is well known that specific cell types are devoted to performing specific functions, but just how many functions might a single cell have? A good place to start is to draw analogies with the technologies that keep our civilizations running: waste management, infrastructure maintenance, transportation, decision making, thermal sensing, pain sensing, threat assessment, communication, replication,[1] etc. Some cells are devoted to locomotion: muscle cells contract or relax when electrical signaled to do so by neurons. We also have on the order of 10^8 cells of a handful of types devoted to light sensing within various wavelengths, or cells that simply come on or turn off when light is present (Kandel et al., 1991). To maintain the mechanical integrity of bones and by extension the bodies they support, osteoblasts and osteoclasts are constantly building and breaking down bone tissue, as a result of chemomechanical signals left behind from mechanical loads (Lemaire et al., 2004).

One question that cell categorization raises is whether cells are discrete and distinct, or whether they should be viewed as spatial and categorical continua. Since all cells within a given organism have a single parent cell (Thomas, 1978; Wolpert, 1991), and indeed all living organisms living today can be traced to a single common ancestor (Dawkins, 2004) there is certainly a spatiotemporal continuum traceable back to both an organism's first cell, as well as the biosphere's origins. However, as we will see in the chapters that follow, it is more convenient to categorize cells on a discrete basis according to their respective functions.

1.1.3 Number of Atoms in a Cell

A human body with a mass of 50 kg has an atomic mass of 3×10^{28}. This is thirty thousand trillion trillion nucleons: 28% is carbon,

[1]Replication of an organism involves the physical transfer of genes by way of molecules called chromosomes, whereas replication of civilization is conducted through memes, which are transferred via numerous media (Dawkins, 1976).

10% is hydrogen, 57% is oxygen, 2.4% is nitrogen, 1.6% is calcium, and 0.85% is phosphorus, CHON+P (Shubin, 2013). Of the 90 naturally occurring elements, life uses 22 of them as building blocks with S, Na, K, Cl, Mg, Si, Fe, Zn, Cu, I, Mn, F, Cr, Se, Mo, and Co comprising the 0.66% balance. Taking and order-of-magnitude approach, if the approximate volume of a human is 0.1 m^3, and a typical cell has a nominal diameter of 10 μm, this yields one hundred trillion cells per body, a number comparable to the Gross World Product. Thus, an average cell has about fifty trillion atoms. The majority of these comprise large molecular complexes such as protein, membranes, DNA, but some are free ions such as sodium, calcium, and potassium.

1.1.4 Number and Organization of Molecules in a Cell

The number of molecules in a single cell is likely to be two to four orders of magnitude lower than the number of atoms. For example, a molecule such as myosin with two heavy chains consisting of about 2000 amino acids each, and four light chains with atomic masses of 20 and 17 kDa has an overall atomic mass of approximately 340 kDa. So if every atom in a cell were part of a myosin molecule, there would be *only* 100,000,000 or so molecules per cell. If, on the other hand, our cell were entirely water, (18 Da), there would be around two trillion molecules per cell.

Other substantial contributors to cell mass are the various membranous structures such as the cell membrane, the nuclear membrane, and the membranes of various cellular complexes such as Golgi apparatus, endoplasmic reticulum, and mitochondria. For example, 1,2-dioleoyl-sn-glycero-3-phosphocholine (DOPC), the molecule that comprises cell membranes, has an atomic mass of 256 Da. If we assume that 5% of the cell volume is membrane, and that the average mass of a phospholipid is approximately 250 Da, then there are roughly 8 billion phospholipid molecules in a cell. Since they are typically found in a 4 nm-thick bilayer, and estimating that the cell volume is approximately 1 pl, the total membrane volume is thus approximately 0.05 pl resulting in a total area of all membranes within a single cell of approximately 10,000 μm^2, or about the cross-sectional area of a single hair.

1.1.5 Mechanical Components of a Cell

Perhaps the most prominent "mechanical" components of cell are the proteins that comprise the cytoskeleton. These are the components that facilitate transport and give the cell its structure as in addition to serving as substrates for numerous protein activities. The combined surface area of these lattice-like cytoskeletal proteins is approximately 250 times that of a cell's membrane itself (Gershon et al., 1985). Among the largest structures seen within the cell are the microtubules. Comprised of dimerized globular proteins, microtubules are 24 nm diameter compression-bearing structures that are found primarily in the nucleus as well as in axons. In fact, it is microtubules that students typically see for the first time under the microscope in their science labs (Fig. 1.1a,b). Under greater magnification, individual microtubules can be resolved (Fig. 1.1c,d).

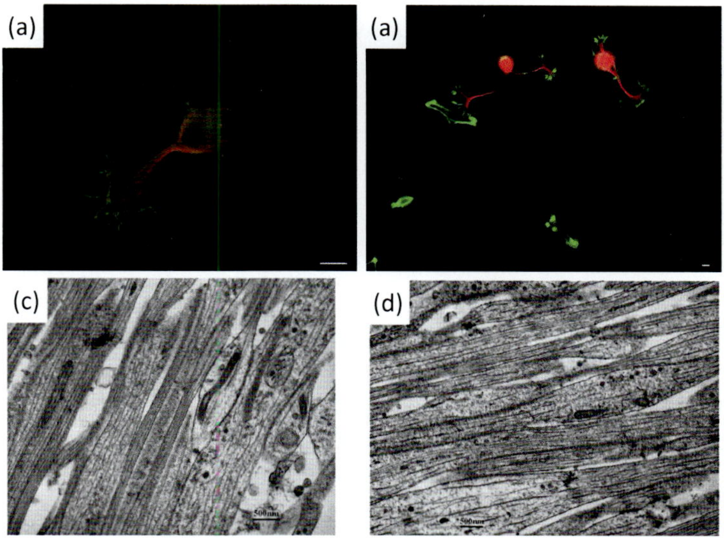

Figure 1.1 (a and b) Rat sympathetic neurons from superior cervical ganglion taken with an inverted Zeiss Axiovert 200M light microscope. Red indicates B3-tubulin and green is phalloidin-stained actin. Scale bar = 10 μm. Imaging by Olga Kahn. (c and d) TEM images of rat superior cervical ganglion transplant culture. Scale bar = 500 nm. (Baas and Black, 1990).

Actin, with a molecular weight of 42 kDa and a nominal diameter of 4–5 nm (EH and R, 1984; Holmes et al., 1990; Lepault et al., 1994; Z et al., 2000) is primarily responsible for bearing tension as can be seen at the periphery of the neurons in Fig. 1.1a,b but can also bear compression when configured in a lattice structure as can be seen in the interior of the neuron in Fig. 1.1a.

Myosin, with its dual heads, acts as a motor, while integrins and cadherins act as dynamic anchors. These struts, motors, and anchors act as a coordinated mechanical network, informed by concentrations, locations, and physiochemical states of a multitude of other molecules to maintain communication between the numerous sensing proteins of the cell membrane and nuclear membrane.

When 10 to 100,000 individual atoms bind together into a large molecular structure, they can begin to perform some remarkable feats. Genomicists categorize proteins into several categories according to their function, i.e., (Chen et al., 2009). In addition to the proteins mentioned above, two other prevalent structural proteins are collagen and keratin. These are found almost exclusively outside the cell. There are also a myriad of motor proteins. In addition to myosin, the two proteins dynein and kinesin are responsible for transporting subcellular structures such as mitochondria and vesicles. While not exclusively structural, channel proteins, which are bound within in the cell membrane, are responsible for regulating the passage of sodium, potassium, calcium, chloride, and other ions for the purpose of controlling the flow of information both within single cells and among cellular networks. Examples of signaling proteins include insulin and testosterone, both of which are found free in the bloodstream and are responsible for cell–cell signaling over the scale of the entire organism. Another signaling protein is rhodopsin, which is an integral (membrane bound) G-protein responsible for transducing light in the retina. Rather than being responsible for standing guard at a particular location in or around cells, these proteins are deployed to regulate or coordinate activities via the bloodstream. Perhaps the most intriguing of all are the replication proteins, e.g., transcriptase, DNA-polymerase, etc. These are typically found in complexes of up to a dozen and literally bind, bend, and break the bonds of life (Ahern, 2013; Barry, 2013).

While we will not likely know for decades the precise mechanisms by which a single cell "decides" how to budget its protein manufacturing and organization efforts in order to maintain its vitality for a prescribed duration, we are able to make a few general observations about individual protein expression profiles for a given cell type, as well as a general picture of the spatial organization of various proteins throughout the cytoplasm, cell membrane and extracellular matrix.

A primary challenge in giving a realistic picture of exactly where and when one might be expected to find an individual protein within a given cell is that there is no deterministic model (Berg, 1999; Kuhn et al., 2011). Another challenge of building an all-inclusive vision of protein distribution within individual cell types is that the 10^7 species on the planet have a combined total of 10^{31} cells. As we saw above, they can be roughly categorized into phenotypes, but it is arguable that our cells are as unique as we are.

Another confounding issue in developing a comprehensive model of cell mechanics is that there are efforts under way to create artificial living systems (Benner, 2010). To confound matters even further, we are now capable of synthesizing virtually any protein sequence up to lengths of several hundred individual amino acid molecules through commercial means (Calhoun and Swartz, 2005; Liszewski, 2010; Araya and Fowler, 2011; Fortenberry et al., 2011). So thus, within a design space of 20^n, where n is the number of amino acids in a single protein, we see that for proteins, which have lengths ranging from 10 to over 1,000, there are $\sum_{i=10}^{1,000} 20^i$ potential proteins. Humans use 20,000 to 30,000 of these.

With this in mind, let us take a moment in this first chapter to examine just two of the numerous proteins found in nature: the protein coat of the bacteriophage T4 virus and the protein that comprises vacuoles in cyanobacteria.

The T4 virus has been fully sequenced and has a DNA strand length of 168,903 nucleotides, and 278 genes. With a volume of approximately 500 zL, it is capable of infecting bacteria by attaching itself to the membrane where it injects its DNA into the cytoplasm using mechanism (Rossmann et al., 2004). Once replication begins inside an *E. coli*, which has a volume of approximately 600 aL, the cell's own suite of replication proteins manufactures hundreds of copies of the virus, including the proteinaceous coat, until the bacterium literally explodes. It became readily apparent that this was certainly not a cell, and that the outer membrane was clearly not a phospholipid.

There has been recent speculation that viruses are only "structurally separable" from cells and have origins that predate what is commonly known as the LUCA (last universal common/cellular ancestor) (Forterre, 2010). With relatively simple structures, viruses replicate quickly inside their host cells: a single virus can have a lifetime that can last for only a few milliseconds or potentially up to millennia if left cold and preserved. It is therefore essentially impossible to "kill" a virus. While we have essentially eliminated the smallpox virus from the human population (Tognotti, 2010), there are still copies of the smallpox virus in laboratories controlled by the CDC. Some of the first electron micrographs of viruses emerged in the 1980s (Adrian et al., 1984) of viruses such as adenovirus type 2, T4 bacteriophage, Semliki Forest virus, and CbK bacteriophage and it continues to be valuable tool (Roingeard, 2008).

The second "simple" class of proteins of interest we will discuss in this chapter includes the vacuole-forming protein of the cyanobacterium *Trichodesmium erythraeum* as well as similar proteins in other single-celled organisms (Walsby, 1972) is a good place to begin to understand the important role that protein mechanics can play in bridging the gap between the molecular world and the cellular world (Sandh et al., 2012). As with all other proteins, the instructions for specifying the sequence of amino acids to be assembled lies within the DNA, in this case at six or seven predicted locations: YP_722017.1, YP_722018.1, YP_722019.1, YP_722020.1, YP_722025.1, YP_722027.1, and YP_722029.1. Of these, YP_722027.1 is the longest, consisting of 153 residues, and a predicted molecular weight of 16.84 kDa, while the others, collectively referred to as GvpA, for "gas vesicle protein A" have molecular weights ranging from 6.23 kDa to 11.27 kDa. The FASTA format for Gas Vesicle K is

1 **m**s**l**enspees **liv**p**i**dksks npe**a**g**l**ap**ll** **l**t**vi**e**ll**re**l**
mq**a**q**vi**rrmd **a**g**il**sdeq**l**e
61 r**aa**eg**l**rq**l**e eq**vi**k**l**ck**vf** d**i**pted**l**n**l**d **l**ge**i**gt**ll**pk
sge**yy**pgeks enps**vl**e**ll**d
121 r**il**ntg**vvl**d gt**v**d**l**g**la**e**l** d**li**h**a**r**l**r**lv** **l**t**a**

whereas the sequence for YP_722029.1 is

1 **m**kts**a**n**ia**ks **a**ggds**la**d**vl** er**vl**dkg**ivi** **a**gd**i**s**v**s**ia**s
te**ll**n**i**k**i**r**l** . ss**v**er**a**ke
61 **i**g**i**n**ww**esdp s**l**ssqnns**lv** q**v**nqk**ll**er**v** **a**s**l**ese**i**e**al**
k**y**sq

Hydrophobic residues have been shaded to emphasize the fact that each of the amino acids in the sequence has physical properties that may be used to predict structure. For example, the publically available program SABLE (Solvent AccessiBiLitiEs and secondary structures prediction) produces the structure for YP_722029.1 as shown in Fig. 1.2.

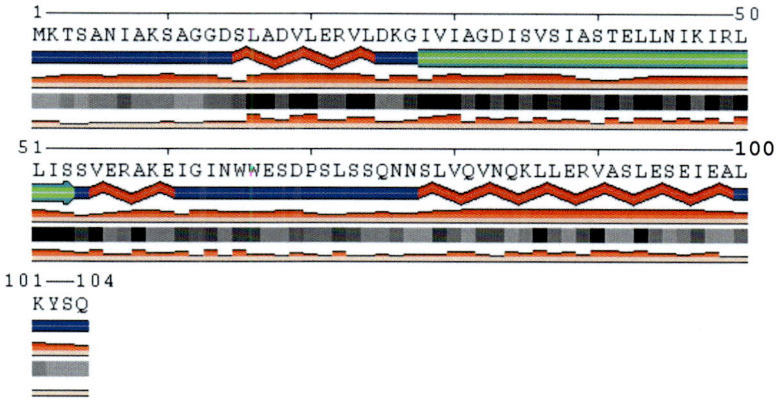

Figure 1.2 Secondary structure plot of vacuole-forming protein YP_722029.1, where the blue is coil structure, the red is α-helix, and the green is β-strand. The line below that represents the hydrophobicity (thicker line = more hydrophobic) the gray scale represents the solvent accessibility with black being more buried and white being more exposed, and the bottom line represents the level of confidence of prediction (thicker line = more confident).

After translation has occurred, the next major step is what is known as protein folding or protein self-assembly. For review, see Hingorani and Gierasch (2014). Essentially what happens during protein folding is the protein "finding" a local energy minimum, which is a superimposed set of five orthogonal energy fields (Zeiger and Layton, 2008).

In mechanical terms, three of these energy fields represent the three primary deformation modes (stretching, bending and torsion) between individual atoms on the protein backbone, whereas short-range interactions and long-range coulomb interactions comprise the other two fields (Compiani and Capriotti, 2013; Dill et al., 1995). It is thus the specific configuration and electrostatic charge of the individual functional groups that define interactions with both nearby functional groups as well as other groups found in the vicinity as a result of folding. These relatively well-understood interactions are used to predict tertiary and quaternary structure. It is indeed the tertiary and quaternary structures that give each protein its unique "personality" and define its role for its organism.

1.2 History of Cell Mechanics

1.2.1 Early Experiments on Measuring Tissue Constitutive Properties

Any student entering the field of biomechanics will almost certainly be exposed to the work of Y. C. Fung, who is commonly regarded as the founder of modern biomechanics. For an example, see his seminal work (Fung, 1993). Fung and his colleagues were able to take the generally messy field of measuring tissue properties and modify well-established material constitutive laws to predict mechanical response to a range of loads. His famous exponential strain constitutive equation for preconditioned soft tissues

$$w = ½[q + c(e^Q - 1)]; \quad q = a_{ijkl}E_{ij}E_{kl}; \quad Q = b_{ijkl}E_{ij}E_{kl} \tag{1.1}$$

uses the full Green–Lagrange strain tensor E_{ij}, and three constitutive constants, two of which are fourth-order tensors: a_{ijkl} (energy per unit volume) and b_{ijkl} (dimensionless), and one of which is a scalar, c (energy per unit volume), to arrive at a strain energy function, w.

For example, his work on the residual stress in arteries (Chuong and Fung, 1986) revealed that the relatively simple act of first sectioning a blood vessel transversely then with a single axial slice along one side would cause the vessel to open revealing that the outer layers are in tension and the inner layers in compression under physiological conditions.

In practice, it is difficult to observe and thus measure the full strain energy tensor, and thus a typical experiment involves loading along a single axis. This method has been applied to other soft tissue, such as peripheral nerve (Abrams et al., 1998; Bora et al., 1980b), which coincidently behaves similar to blood vessels in that it too has a slight internal pressure (Low et al., 1977; Yabuki et al., 1998; Layton, 2003b). In later chapters, we will see the value of constitutive modeling from the molecular scale up to the tissue scale.

1.2.2 ECM Mechanics versus Cell Mechanics

Some of the earliest work where investigators attempted to measure the mechanical properties of soft tissue such as nerve include Bora et al. (1980b), Andreassen et al. (1981), and Beel et al. (1984). Later work includes that of Abrams et al. (1998) and Layton and Sastry, (2004, 2006). If we compare the moduli obtained in many of these experiments, which typically range from 1 to 20 MPa, we see a great contrast to some of the cell mechanics modulus testing that has been occurring lately where individual cell moduli are more in the (1–10 kPa) range (Dokukin et al., 2013; Ren et al., 2013). Thus, as expected it is the extracellular matrix that contributes the majority of the load-bearing in larger scale experiments. This observation is further underscored by more recent cell modulus measurement experiments where the moduli of individual cells; for example, Müller cells (Park and Lee, 2013) have moduli on the order of a few kPa or even a fraction of a kPa (Lu et al., 2013), or isolated aortic valve interstitial cells with moduli on the order of 10 kPa (Liu et al., 2013).

1.2.3 First Experiments and Motivation for Measuring Cell Stiffness

Several of the methods for measuring cell mechanical properties are summarized by Van Vliet et al. (2003). As can be seen from the review, the compliance (stiffness) of the instrument being used must be similar to the sample under investigation. It is also important to keep in mind that even though the modulus of the sample under investigation may be scale independent, the forces involved as the sample size diminishes also diminish. Likewise, since as the linear sample size diminishes, the volumetric sample size diminishes by the cube of the linear dimension. Thus the total strain energy e.g., a_{ijkl} (Eq. 1.1) that the sample can sustain also diminishes by the cube of the linear dimension of the sample.

Some of the most delicate tensile testing experiments include those performed by optical stretching. This pioneering work eliminated physical contact with the cells under investigation and replaced a physical probe with a pair of lasers (Guck et al., 2001). Capable of measuring forces on the order of a few hundred piconewtons, it is suitable for exerting sufficient force to cause optically detectable deformation at the cellular scale. For smaller forces and deformations, non-optical methods are required.

Atomic force microscopy has the ability to resolve deformations smaller than any other mechanical characterization method (Radmacher, 2002). While it cannot typically reach the smaller forces that optical trapping can (Ashkin and Dziedzic, 1987), it does have the ability to perform two-dimensional force mapping, a technique particularly suitable to mapping material anisotropy of cells (Rotsch and Radmacher, 2000).

1.3 Outline of the Book

The remainder of the book's chapters are arranged according to the scales they represent: molecular–cellular–tissue. Chapter 2 summarizes much of Alan Hunt's theory and research into the role that physics plays in understanding how individual molecules may be treated like individual discrete machines. Chapter 3 delves into the mechanics of DNA as well as the nucleus itself, highlighting many of the key molecules in regulating nuclear function. Chapter 4 gives an overview of atomic force microscopy, highlighting some of the structures it is capable of imaging. Chapter 5 takes a detailed look at many of the subtle issues of cell–substrate interactions, focusing on proliferation and motility rates. Chapter 6 provides an in-depth look at the key molecules in axonal transport, and finally, Chapter 7 captures many of the key molecular features of coronary disease.

Problems

1.1 ***Predicting molecular distributions.*** (a) Write a function to predict how many carbon atoms are in a typical organism. Make assumptions on water content, protein content, carbohydrate content and lipid content. (b) Write a function to predict how many cells are in an organism.

1.2 ***How much DNA?*** Make a prediction for the total length of all the existing DNA on the planet. How unique are some regions? How common are others?

1.3 ***Going viral.*** Do some independent investigation into the origins of viruses. Do they evolve separately from their hosts, on in tandem with their hosts? What are their ultimate origins? How likely is it for viral DNA to make it into a species' genome?

1.4 ***The first cells.*** Do some independent research and see when scientists think the first true cells evolved. What was the earth's climate like then? What was the main mode of metabolism.

1.5 ***Going big.*** Do some independent research and see when scientists think that multicellular life first emerged. How many cell types would there be in a single organism? How many total cells?

Chapter 2

Mechanics of Single Molecules and Single Proteins

Alan Hunt, Blake Charlebois, and Seth McCubbin

Biomedical Engineering Department, The University of Michigan, Ann Arbor, MI 48109, USA

charleb@umich.edu

2.1 Macromolecules, Small Molecules, and Machines: How Are They Alike? How Do They Differ?

We begin this chapter by first classifying objects as either macroscopic or microscopic to determine how to best analyze their behavior. This distinction is not trivial. In fact, it has launched the field of nanoscale biomechanics, which bridges the scale between molecules, the vast majority of which have a linear dimension of less than 100 nm, and cells, which typically have a linear dimension of more than 1 μm. Examples of small molecules include water (H_2O), benzene (C_6H_6), acetic acid (CH_3COOH), and adenosine triphosphate (ATP) ($C_{10}H_{16}N_5O_{13}P_3$). Usually we consider the behavior of these on purely chemical terms

Molecular and Cellular Biomechanics
Edited by Bradley E. Layton

ISBN 978-981-4316-83-5 (Hardcover), 978-981-4613-44-6 (eBook)
www.panstanford.com

(thermodynamics, formation and breaking bonds). And typically, the thermochemical behavior of these molecules is measured in samples containing molar quantities of these molecules, i.e., $>10^{23}$ molecules in solution. Consider hydrogen and oxygen reacting to produce water and heat:

$$2H_2 + O_2 \rightarrow 2H_2O + \text{heat}. \quad (2.1)$$

2.1.1 Macroscale vs. Microscale

Likewise, when we examine the behavior of large mechanical structures such as automobile engines we examine the behavior of such systems by considering thermal expansions, mechanics of gears and levers. These systems, which might contain tens to hundreds of times the mass used in a benchtop experiment, are also governed by the same laws that govern the reaction of equation (2.1). The overall mechanics of any combustion engine may be outlined as follows:

combustion → work by expanding gases → levers move, gears turn, wheels turn → car moves

So essentially, the vast majority of our technologies are chemotrophic rather than phototropic: we convert paleopetrochemical photosynthetic energy into kinetic and ultimately thermal energy at a rate ten million times greater than the rate at which it was stored. That said, all species that do not employ technology, i.e., every species but one, extract energy from the environment that has only been recently stored. It is thus of primary importance that the biotechnologist understands the fundamentals of the molecules that allow for this conversion in the absence of technology.

2.1.2 Macromolecules Behave Like Macroscale Machines

Biological macromolecules are of a scale that allows them to exhibit behaviors falling between the well-described statistical mechanics of small chemicals, and the similarly well-characterized mechanical behaviors of macroscopic objects. The traditional chemist's perspective is in many ways appropriate for describing macromolecules; yet, like macroscopic objects, individual macromolecules are large enough to undergo stable changes in their internal conformation, and thus may contain elements that behave, for example, as springs (elastin), levers (myosin), or driveshafts (flagellum, ATP synthase). This allows macromolecules to function as scaffolds and rails (microtubules (MTs)), or even motors (kinesin) and strain gauges (integrin) (Howard, 2001). Despite this, the mechanical properties of most biological macromolecules have yet to receive full attention; most research either focuses on biochemistry (study of chemical processes in biological organisms) or molecular genetics (study of gene structure and function at the molecular level). This is not surprising; until the advent of technologies such as atomic force microscopy (Binnig et al., 1986), no tools existed for precise mechanical manipulation at the sub-micrometer scale.

As a consequence of their extraordinarily diminutive size, biological nanomachines often exhibit behaviors that are unexpected or contrary to intuition. For example, some calcium pumps in a cell membrane use energy released by ATP hydrolysis to pump calcium across the membrane. However, if the calcium gradient is large enough, these pumps can be driven backward so that they synthesize ATP rather than hydrolyzing it (Hasselbach, 1978). It would be a very unusual car that could synthesize petrol and O_2 when its wheels were forced to turn backwards! We will return to this conundrum at the end of the next section.

So let us now explore the nanomachines that allow cells to literally work. As formalized to a degree by Schrödinger, the second law can be violated locally, but not universally. This notion is given a mathematical treatment by Lebowitz (1999), where he invokes the work of Boltzmann to discuss how the low-entropy conditions of the early universe have gotten us to the relatively more entropic universe that we observe today. Just taking the example of how ATP and myosin interact, we see that this molecule (ATP), which constitutes only about 8 oz of an average human's body mass, or about 0.5 mol per person, transforms the myosin molecule between its two states, 10^3 to 10^9 times per second (Yu et al., 2007), or likely in the neighborhood of 3×10^{29} times per person per second. Since one "person power" is approximately 100 W, this means that the ATP-myosin interaction has a power of about 10^{-27} W, or that ATP has a power density of about 200 W per pound.

An organism's replication of cells, most of which are identical, i.e., red blood cells, muscle cells, do not significantly increase its entropy, and indeed it is the high level of redundancy that reduces an organism's probability of falling victim to entropization. It is also important to note that all of the chemical energy entering an organism is then converted to other forms, such as electrical, mechanical, or other chemical forms, but that ultimately all of the energy becomes thermal and is discharged to the environment. Thus, to understand the functioning of cellular and subcellular structures, it is critical to consider the triad of mechanics, physics, and chemistry.

2.2 Thermal Energy, Equipartition, and the Boltzmann Distribution

2.2.1 Thermal Energy and Thermal Forces

Thermal and viscous forces dominate molecular mechanical phenomena. When a molecular motor exerts force against a filament or when a RNA polymerase moves along DNA as it performs transcription it must contend with collisions with surrounding solvent and other molecules (Fig. 2.1). Furthermore, viscous effects dominate over inertial effects (i.e., the Reynolds number is low, perhaps 10^{-8}), so when applying Newton's second law, mass can often be neglected (Purcell, 1977). Furthermore, relaxation of stretched proteins is like that of a highly over-damped spring (Howard, 2001).

The "entropic spring" provides an illustrative example of thermal effects, in that it models the work done on a system as it is moved away from its equilibrium (high entropy) state, and it can provide the basis for understanding seemingly different molecular processes. Consider a rope attached to a wall. In our macroscopic world, I can easily make the rope very straight (Fig. 2.2) by pulling the free end; however, if a crowd of people arrives, with each person standing and pulling randomly at some point along the rope, I must exert a larger force and thus expend more energy to keep the rope straight. In other words, it is "easier" (less power is required) to allow the crowd to induce a range of spatial configuration for the rope (Fig. 2.2B,C) than it is for me to continue to try to maintain the preferred straight conformation. The crowd of people represents of course the solvent, and its associated molecules that collide with the entropic spring (my rope) or with the molecular motor (me holding the rope).

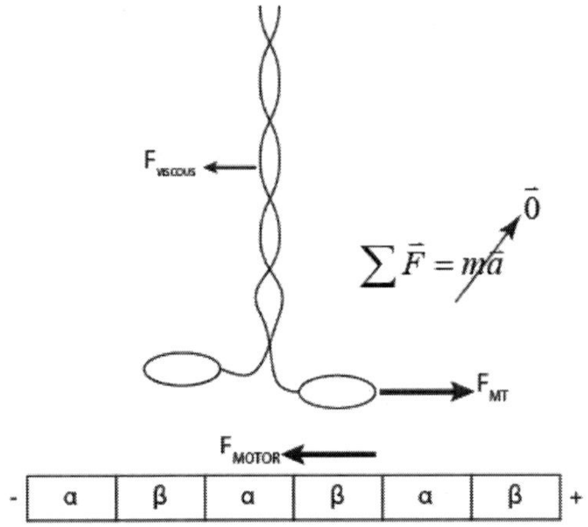

Figure 2.1 Kinesin-1 walks along a microtubule protofilament (see Fig. 2.5) toward its + end, away from the nucleus. Exchange of energy between kinesin and the surroundings due to collisions with solvent molecules (thermal forces, not shown) ensures that the kinetic and potential energy of different portions of kinesin (or different deformation modes) is always changing. The mass, m, of kinesin is small enough that the inertial term $m\bar{a}$ is much smaller in magnitude than the viscous drag or the reaction force of the microtubule on kinesin.

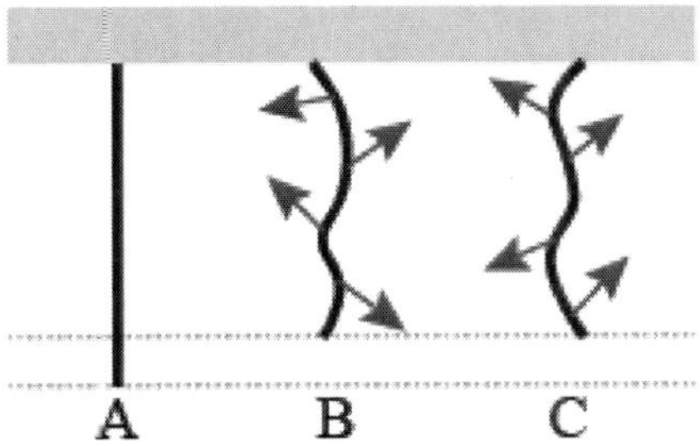

Figure 2.2 An entropic spring. (A) A large force is applied downward in order to maintain a large end-to-end distance. (B, C) A smaller force is applied, and thermal forces cause the entropic spring to take on a range of configurations, reducing the end-to-end distance.

This type of model can be applied to various biological macromolecules, and similar thinking can be extended to other entropic phenomena: for instance, an N_2-permeable membrane would require force to keep all the N_2 molecules on one side of the membrane. A chemical interconversion between two species provides a related example. Suppose this interconversion involves no net change in internal energy, pressure, or volume (i.e., no enthalpy change), and suppose we begin with a much greater concentration of one species than the other. It would require work to maintain this concentration difference against random interconversion due to thermal forces. This concept will arise again below, in the Section 2.5, where we will find that working against thermal forces requires release of heat into the surroundings.

We return to the observation that some biological motors can be operated in reverse while, for instance, combustion engines, and indeed organisms, cannot. Consider two factors. First, biological motors almost always operate near thermodynamic equilibrium, while macroscopic motors generally do not. Thus, the heat lost during a chemical transition in a biological motor can easily be absorbed from the environment to support reversal. This is particularly important for the transitions that are not tightly coupled to force generation (and thus cannot derive the extra energy from the applied force). A second factor is that macroscopic motors involve, for instance, expansion of gases containing huge numbers of molecules. Consider that Newton's laws are time-reversible, so that if a movie is taken of a simple system of a few colliding particles, this movie will seem realistic whether played forward or backward. Why does this not extend to more complicated systems? In short, the forward direction corresponds to an entropy increase from an improbable state toward a set of more probable states. For instance, if we concentrate all the molecules in the room into one corner they will expand to fill the room. Assuming they have elastic collisions with the walls, we could reverse their momenta at a given moment to have them concentrate in the corner again, but it is improbable that they would spontaneously acquire a set of positions and momenta that would cause them to concentrate in this way. In a system with far fewer particles such spontaneous concentration would be more probable. For further discussion, see Chapter 46, "Ratchet and Pawl," in Feynman et al. (1963). We now turn our attention to determining when thermal forces are important by quantifying their magnitudes.

2.2.2 The Equipartition Theorem

The equipartition theorem may seem at first like an obscure statement of little utility. However, when it comes to estimating thermally induced vibrations, an important source of experimental "noise," it is tremendously useful. A common definition for the equipartition theorem is

> *"At equilibrium, for systems having an internal energy (kinetic or potential) that is proportional to the square of a position or speed, the mean of that energy is one-half Boltzmann's constant times temperature or,* E = ½k$_B$T."

More precisely, each degree of freedom that enters quadratically in the expression for energy contributes mean energy $\tfrac{1}{2}k_BT$. Thus, it may be applied to the three fundamental deformation modes (tension/compression, torsion, and bending) and to kinetic energy. Here, T is the absolute temperature; at absolute zero, thermal effects vanish. (Note that in equations, entropy often appears in a term in which it is multiplied by T, such that entropic effects vanish at absolute zero.) Boltzmann's constant is equal to the universal gas constant divided by Avogadro's number.[1] At biologically relevant temperatures of say 300 K, $\tfrac{1}{2}k_BT \approx 4\,\mathrm{pN\cdot nm}$, which is approximately the energy (work) that a single kinesin-1 motor exerts (performs) while transporting a vesicle 1 nm against about half the force at which it would stall.

What are the practical implications of the equipartition theorem? Kinetic energy of a particle at both the macro and molecular scale is proportional to the square of speed. For a particle with x-velocity v_x and mass m, the energy is $\frac{1}{2}mv_x^2$, so the equipartition theorem gives $\tfrac{1}{2}k_BT = \left\langle \frac{1}{2}mv_x^2 \right\rangle = \frac{1}{2}m\left\langle v_x^2 \right\rangle$ and root mean-squared (rms) x-speed $\sqrt{\left\langle v_x^2 \right\rangle} = \frac{k_BT}{m}$. For a particle moving in

[1]When we work in units of 6.02×10^{23} molecules, we use the gas constant. When we work in units of 1, we use Boltzmann's constant.

three dimensions at speed v, the mean kinetic energy is $\frac{1}{2}m\langle v^2 \rangle = \frac{3}{2}k_B T$.[2]

But what are the practical implications for molecular mechanics? The energy stored in a stretched protein can often be assumed to be proportional to the square of the stretch length, $E = \frac{1}{2}\kappa x^2$, where the spring, in this case the protein, is modeled as a Hookean spring with spring constant κ. Using the definition of the equipartition theorem, and setting the thermal energy equal to the potential energy, we can estimate that thermal forces cause rms stretching and thus forces according to $F = \sqrt{k_B T/\kappa}$. For instance, if one head of kinesin is bound to a microtubule (Fig. 2.3a), we can determine how far (on average) the other head will typically reach due to thermal energy.

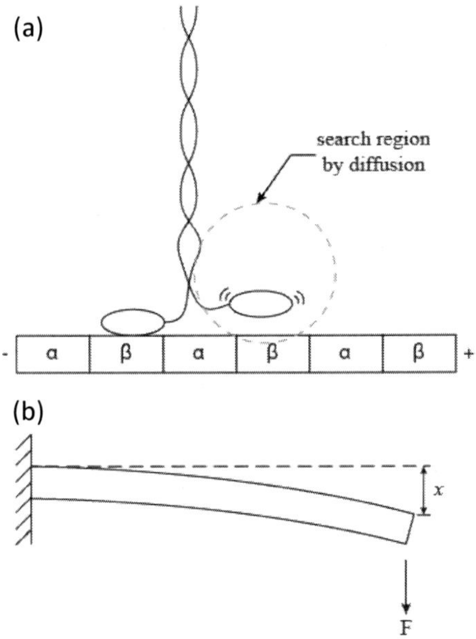

Figure 2.3 (a) The tethered head of kinesin can be modeled as being restrained by a spring, and the equipartition theorem allows us to predict how far it can diffuse as it searches for its binding site on β-tubulin. (b) A microtubule can be modeled as a flexible beam with a spring constant associated with bending, which allows its thermal bending to be predicted.

[2] $\frac{1}{2}m\langle v^2 \rangle = \frac{1}{2}m\langle v_x^2 + v_y^2 + v_z^2 \rangle = \frac{1}{2}m\langle v_x^2 \rangle + \frac{1}{2}m\langle v_y^2 \rangle + \frac{1}{2}m\langle v_z^2 \rangle = \frac{3}{2}k_B T.$

We can also estimate the torsional stiffness, K, of a kinesin motor by observing the variance in angles of an attached microtubule (Hunt and Howard, 1993). A variance lower than that expected for free diffusion tells us that kinesin has a nonzero torsional stiffness. Using the equipartition theorem, we can relate the thermal energy to the torsional energy such that

$$\frac{1}{2} K \, \text{variance} = \frac{1}{2} k_B T \quad \text{or} \quad K = \frac{k_B T}{\text{variance}}. \tag{2.2}$$

Similarly, we can apply the equipartition theorem to look at microtubule bending induced by thermal fluctuations. Consider a microtubule modeled as a cantilevered beam (Fig. 2.3B). The bending energy is defined as

$$\tfrac{1}{2}\kappa x^2 = \tfrac{1}{2} k_B T, \tag{2.3}$$

where κ in this case represents the stiffness of the cantilever. This analysis can be extended to account for all bending modes (Gittes et al., 1993).

Thermal forces are crucial to molecular and cellular biomechanics because $k_B T$ is on the order of the energy from ATP or GTP that drives protein movement, and also because it is sufficient energy to cause some significant conformational changes in proteins.

To complete our analogy with the power of a macroscale engine, we estimate the power density of a kinesin molecule. With a mass of approximately 250 kDa (4.2×10^{-22} kg), and moving along a microtubule at approximately 1 μm s^{-1} (125 steps s^{-1}), kinesin expends energy at a rate of 3250 $k_B T$ s^{-1} or approximately 1.4×10^{-17} W. This is equivalent to 34 W g^{-1} (45 hp kg^{-1}). This rate is not constant: at half of kinesin's stall force it expends 28 pN nm per step or 6.5 $k_B T$ per step (8.5 W g^{-1}). Regardless, the power density of a Formula One engine with a mass of 95 kg and a power rating of 750 kW (1000 hp) has a power density of about 8 W/g. In other words, a macroscale engine, which has been engineered to operate at a temperature where material thermal limits become a factor, has yet to achieve the power density of one of nature's most intriguing molecular motors.

2.2.3 The Boltzmann Distribution

The Boltzmann distribution, from which the equipartition theorem can be derived, provides a more flexible quantification of thermal effects, resulting in numerous implications. Suppose a particle can be in any of a set of states [i], and that when it is in state i it has energy U_i. For example, in (Fig. 2.4), the particle could be an He atom taking on continuous values of kinetic energy, a Na^+ ion taking on continuous values of electric potential energy, or a H_2 molecule that can take on only a discrete set of vibrational energy levels. Boltzmann's law states that at equilibrium, the probability p_i that a particle will be in state i is proportional to e^{-U_i/k_BT}.

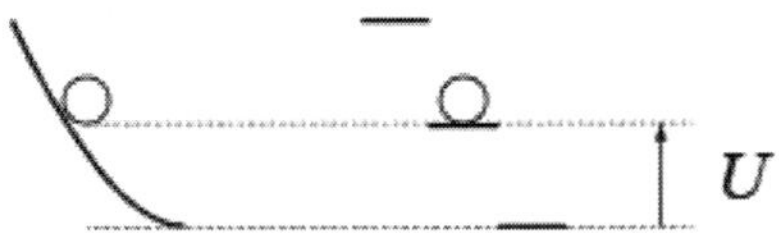

Figure 2.4 The physics of the system under study determines whether continuous (left) or discrete (right) energy levels must be used. If the system can only take on discrete energy levels, but the distance between these levels is much smaller than k_BT, then a continuum approximation may suffice (Feynman et al., 1963).

In our macroscopic world, objects tend to minimize their potential energy—for instance, a ball rolls down a hill, finding itself at "the bottom." By contrast, in the molecular world, when the potential energy difference between two states is comparable to k_BT, thermal forces have a high probability of pushing some particles into higher potential energy states: the macroscale equivalent of wind blowing our ball around in its valley. Consider a particle that can have two potential energy states $U_2 > U_1$. The probability of finding the particle in state $i = 2$ relative to the probability of finding it in $i = 1$ is

$$\frac{p_2}{p_1} = \exp\left(-\frac{U_2 - U_1}{k_BT}\right). \tag{2.4}$$

If the energy difference is large enough that $U_2 - U_1 >> k_BT$, as would be the case with a macroscopic ball rolling down a hill, we have $p_2 \approx 0$, and the particle will almost certainly be found in the lower energy state because thermal energy is insufficient to move it to the higher energy state. In the opposite case, one might expect the opposite result: that if thermal energy is large relative to the energy difference, then thermal forces would keep the particle in the higher energy state, but more careful inspection of (1) reveals that this would not be the case. With $U_2 - U_1 << k_BT$, we have $p_2 \approx p_1$. The particle has enough thermal energy that it does not "feel" the small difference between the two energy states, and thus fluctuates between the two, spending time in each in such a way that $t_2/t_1 \approx p_2/p_1$.

Boltzmann's law can be obtained from the Boltzmann distribution,

$$p_i = \frac{e^{-U_i/k_BT}}{\sum_j e^{-U_j/k_BT}}, \tag{2.5}$$

where the denominator is a normalization constant that ensures the probabilities sum to 1. Here we have assumed discrete energy levels, U_i, but the distribution can also be applied to a particle in a continuous energy well, such as the gas molecules in our atmosphere (Feynman et al., 1963; Howard, 2001). The use of continuous or discrete energy levels is dictated by the physics of the system being studied (Fig. 2.4).

As a biological example we follow Hill (1986, 1987): consider the polymerization of a microtubule from heterodimeric subunits (Fig. 2.5). For simplicity, we ignore the lattice seam, and assume no nucleotide hydrolysis in the subunits near the tip. To construct a model of polymerization, we must consider association and dissociation rates. A subunit with two lateral neighbors (i.e., in the horizontal direction in (Fig. 2.5b,c) will be more stable than one with one or zero, so there are three dissociation rates. A simplification that must be examined is the assumption that microtubule tips are always blunt (Fig. 2.5b), so that all tip subunits except one have two stabilizing lateral neighbors.

To use the Boltzmann distribution to address this assumption, we must quantify the free energy difference between tip configurations, which depends on the number of "unused" lateral bonds. Given the free energy of lateral bond formation, if we enumerate the possible configurations of subunits at the tip, we can use the Boltzmann distribution to obtain the relative probabilities of these configurations at equilibrium (i.e., at the critical concentration and in the absence of GTP hydrolysis). From this we conclude that the tip is blunt only a fraction of the time, in which case we must consider different dissociation rates depending on tip configuration. In fact, experimental observations of tip structure and estimates of the free energy of lateral bonds indicate that tips should not be assumed to be blunt (Chretien et al., 1995; Gardner et al., 2011).

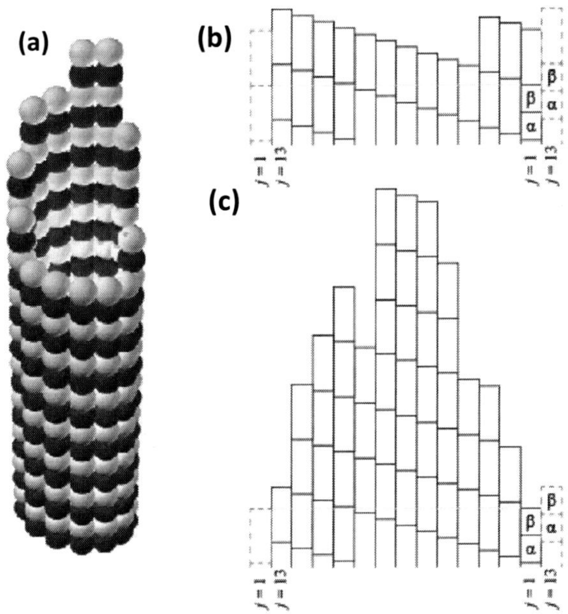

Figure 2.5 (a) Microtubules are formed by polymerization from heterodimers of α-(black) and β-(white) tubulin into a tube. Heterodimers (subunits) bound in the longitudinal (vertical) direction are termed protofilaments, indexed by j in b and c. The seam between protofilaments 1 and 13 is neglected. (b) A blunt tip configuration in which all subunits but one have 2 lateral (horizontal) neighbors. (c) A tip configuration in which some subunits are less stable, having 0 or 1 lateral neighbors.

The broad applicability of Boltzmann statistics is reflected in the appearance of k_BT or RT in diverse equations. For instance, the Einstein–Smoluchowski relation states that the diffusion coefficient of a particle is the ratio of k_BT to the drag coefficient; we can think of diffusion as an opposition of thermal forces that give particles a range of speeds and drag forces that particularly oppose higher speeds. For ions on opposite sides of a semipermeable membrane, the Nernst equation relates the equilibrium concentration difference to the electric potential difference. In chemical kinetics (Fig. 2.6), rate constants are predicted to be proportional to $e^{-\Delta G_a/k_B T}$ where ΔG_a is the activation energy; this reflects the fact that the relative probability of particles acquiring sufficient energy to overcome the activation barrier is proportional to $e^{-\Delta G_a/k_B T}$. A similar line of reasoning yields the law of mass action.

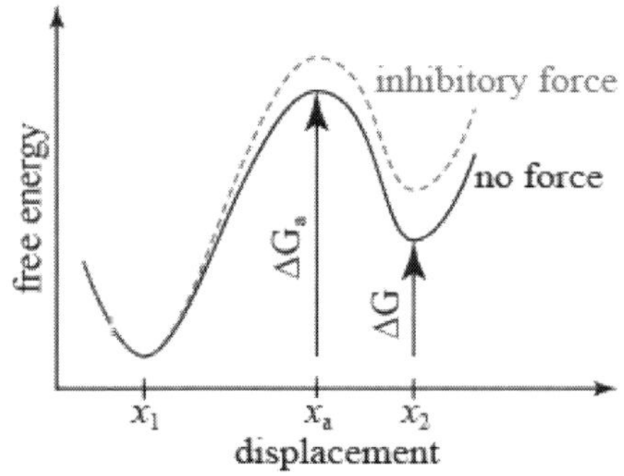

Figure 2.6 The free energy profile of a reaction, including the effect of force. Here the reaction coordinate is assumed to be a displacement from x_1 to x_2 and the intermediate (activated) state occurs at x_a. The inhibitory force F (convention: $F > 0$ inhibitory) increases the required activation energy to $\Delta G_a + F(x_a - x_1)$, affecting reaction rates. It also increases the net free energy change to $\Delta G + F(x_2 - x_1)$, changing the equilibrium constant by a multiplicative factor $e^{-F(x_2 - x_1)/k_B T}$. This is based on the more extensive treatment of Howard (2001).

2.3 Thermal Ratchets: What Are They? A Practical Definition

Examples of mechanical ratchets at the human scale abound from socket sets, to bicycle freewheels, to the winding mechanism in a mechanical wristwatch. In each of these cases, a small effort must be overcome against friction and spring deformation to strongly bias or indeed ensure that motion only occurs in a single direction. On a microscopic scale, thermal effects introduce complications. We begin with non-biological examples.

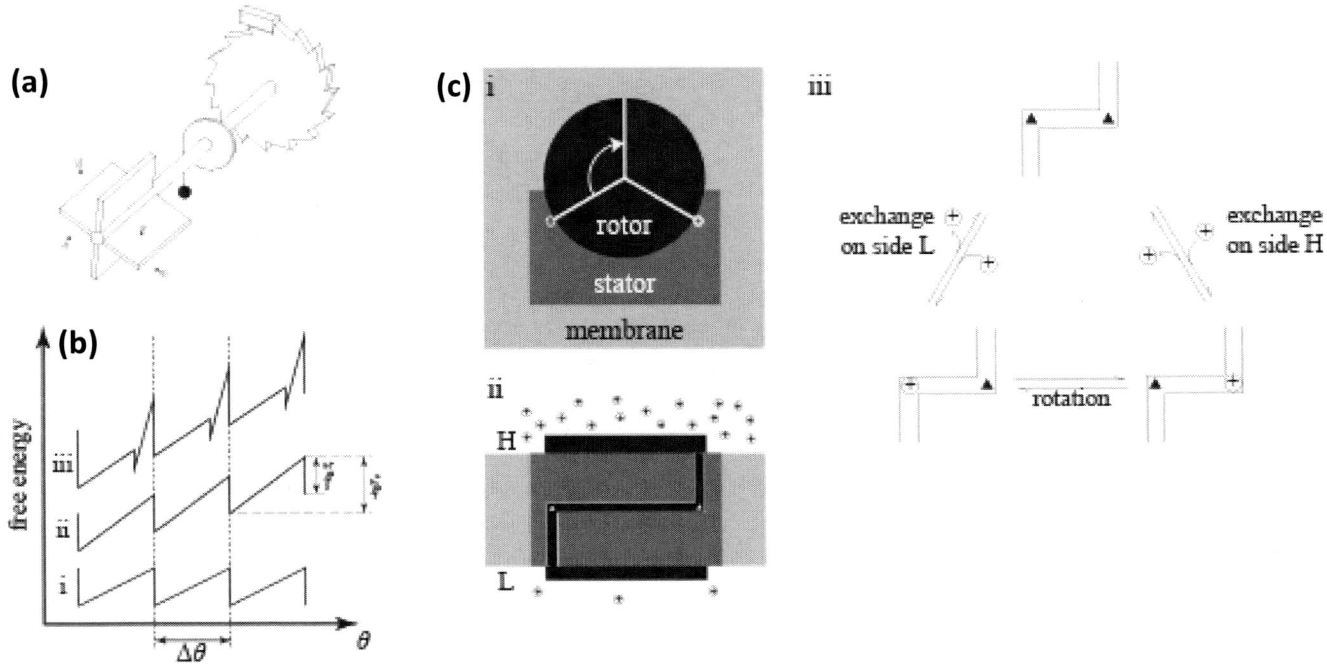

Figure 2.7 (a): In the thermal ratchet, thermal forces bombard the vanes (bottom left), causing the shaft to rotate in random directions. At the opposite end of the device, a pawl rectifies the motion by blocking the ratchet teeth against rotation in the undesired direction. A spring (not shown) must pull the pawl back down again to maintain rectification. However, if

the entire device is at the same temperature, the pawl cannot rectify motion because it too is lifted by thermal forces too frequently. For rigorous treatment, see Feynman et al. (1963). If the ratchet-end of the device is at a lower temperature than the vanes, rectified motion is possible and the device can even do work, in this case lifting a small black sphere. B: Free energy of the device (including the pawl spring and, except in case i, the black sphere) as a function of rotation angle. (i) Neglecting the mass of the black sphere, when the entire device is at the same temperature, rotation through $\pm\Delta\theta$ from one ratchet tooth to the next requires that the energy barrier to rotation (lifting the pawl) be not too much larger than k_BT. Direction of rotation is unbiased, regardless of the shape of the ratchet teeth. (ii) We reintroduce the black sphere and provide an input of energy by (as an example) maintaining the vanes at a temperature T_v, which is greater than that of the ratchet and pawl T_r. The magnitudes of k_BT_v and k_BT_r are chosen such that rotation is rectified. The total energy of the system can increase, here stored in the gravitational potential energy of the black sphere and (cyclically) in the pall spring. (iii) A two-step process for rotation through $\Delta\theta$ in which thermal forces are sufficient to allow the ratchet to overcome the first energy barrier (in either direction) but rarely the second without additional energy input. To make this example more similar to a biological example, we can set $T_v = T_r$ and view rotation of the ratchet as a two-step chemical reaction in which the first reaction occurs by rotational diffusion in the forward and reverse directions while the second reaction requires an energy input (e.g., ATP hydrolysis) to overcome the high activation energy barrier at an appreciable rate. C: A biological thermal ratchet. Several types of membrane-bound rotary motors couple transmembrane ion movement to torque generation. (i) Top view. (ii) Side view (e.g., Elston et al. (1998)). In this example, the stator has a channel (white circles in i, white lines in ii) that allows ions to pass from the high-concentration side of the membrane (H) to the low-concentration side (L). However, to move in the horizontal direction they must be loaded onto a rotor binding site (triangle). In the absence of a bound ion there is a large energy barrier to rotation, but once an ion is bound the rotor can undergo rotational diffusion. Rotation is possible in either direction, but because ion binding is more frequent on side H, the rotor more frequently rotates left in (ii) and clockwise in (i). (iii) Rotor movement can be described in terms of a cyclic chemical reaction. We follow Berry and Berg (1999). Two of the reactions represent the binding and unbinding of ions to and from the rotor, while the third represents rotation. As long as the concentration gradient is maintained, the reaction favors the clockwise direction. Given the rate constant for each reaction, we can compute the kinetics of the motor as a function of the concentration gradient and any torque against which it must work (Berry and Berg, 1999).

2.3.1 Maxwell's Demon

When Maxwell proposed his thought experiment of a "demon" opening a door only to let faster moving particles through thus biasing the direction of energy flow, it was quickly realized that the demon would violate the second law. Similarly, if Maxwell's Demon lifts the pawl of a microscopic thermal ratchet (Fig. 2.7a) when the ratchet's thermal vibrations cause it to move in the desired direction, the second law is violated (assuming all components are at the same temperature).

However, thermal motion *can* be rectified *if* energy is expended (Fig. 2.7b-ii). This is true at the macroscale. Imagine a bicycle bouncing around in the back of a pickup truck. Regardless of how the bike is jostled, the rear wheel (a freewheel device) will only rotate in the forward direction even though the jostling energy from the truck bed is unbiased with respect to whatever configuration the bicycle finds itself. So should we specify movements that have substantial thermal components as thermal ratcheting? No—this would not be meaningful since this would include any chemical transition (two energy wells separated by a transition energy). What if we were to specify that the thermally driven movement is separate from the rectification mechanism? For example, suppose Maxwell's Demon lifts the pawl on a freewheel, but only when it is turning in a preferred direction. In kinetic terms this is equivalent to the existence of multiple transitions involved in generating sustained movements (e.g., a thermally driven movement, and an enthalpic rectification; (Fig. 2.7b-iii), and, at least when the system is not externally loaded, only the transition that rectifies the movement involves a substantial change in energy (potential wells are different height).

As a biological example, to synthesize ATP, the enzyme ATP synthase uses the energy released from protons as they move down an electrochemical gradient, from one side of a membrane to the other (Figs. 2.7c and 2.8). As discussed earlier, this reaction can also be driven in reverse. Given a sufficiently large ATP gradient and relatively low proton gradient, ATP synthase uses energy from ATP hydrolysis to pump protons across the membrane.

As it turns out, Maxwell's Demon has a lot in common with any organism, regardless of size, as organisms ranging in scale from a single cell all the way up to a human are expending energy to tip the odds of thermodynamics in their favor.

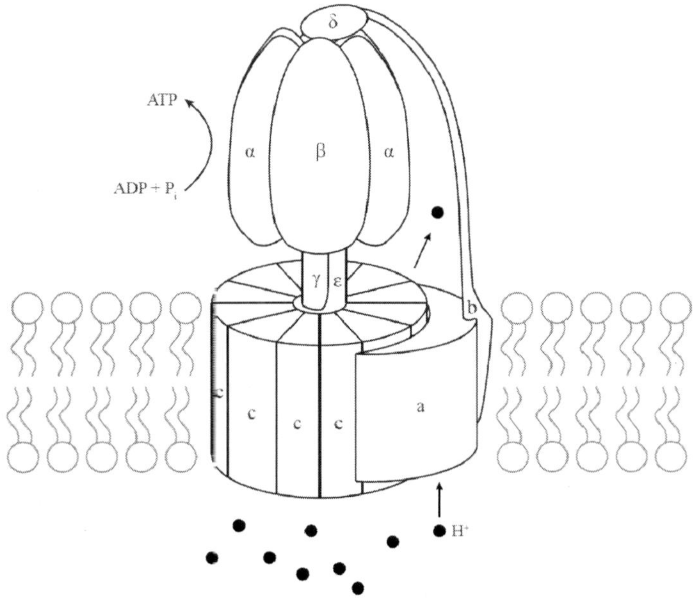

Figure 2.8 ATP synthase synthesizes ATP using the energy obtained as protons move across a membrane. The energy is due to both a concentration difference and an electrical potential difference. The rotor and stator in the membrane (a and c) are similar to Fig. 2.7ci–ii. Work done by the rotor as transmitted through the stalk, γ, is used to generate ATP. This energy drives rotation of the γ stalk, an asymmetric subunit at the center of ATP synthase. γ directly interacts with the $\alpha\beta$ complex, which is kept from rotating by an external stalk, causing conformational changes that combine ADP and P_i into ATP.

2.4 Detailed Balance

The principle of detailed balance states that "at equilibrium, a given elementary process is in equilibrium with (occurs at the same rate as) its reverse process." Examples include elastic collisions, emission, and absorption of radiation, or reversible chemical reactions.

Perhaps the best way to introduce detailed balance is via the following example: consider the force created by a filament polymerizing against a barrier (e.g., actin or microtubule) (Hill, 1981). Using Hill's chemical energy terminology, we provide an example of how this is rewritten with molecular energies so that the relationship between Boltzmann energy and Gibbs energy is clearly seen. The maximum work that can be derived from a chemical transition is defined by the free energy:

$$\Delta G = \Delta G_0 + RT \ln \frac{[\text{products}]}{[\text{reactants}]}. \tag{2.6}$$

For example,

$$\Delta G_{\text{ATP}} = \Delta G_{0,\text{ATP}} + RT \ln \frac{[\text{ADP}][p_i]}{[\text{ATP}]}. \tag{2.7}$$

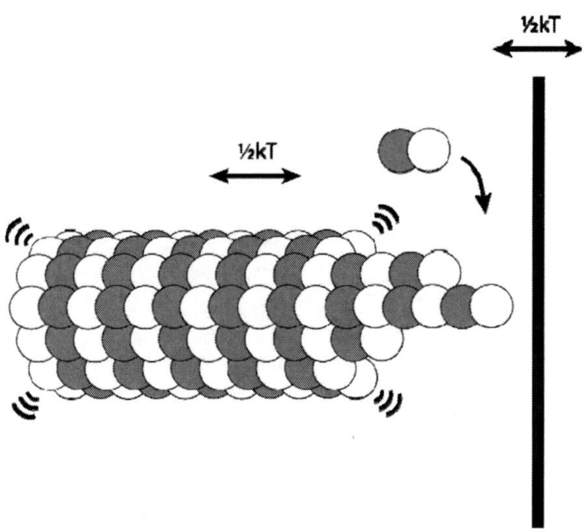

Figure 2.9 Model of tubulin polymerizing against a wall. In this case the polymerizing microtubule and the wall are both subjected to thermal forces and have displacements proportional to $\frac{1}{2}k_BT$, thus allowing for the probability of a tubulin subunit to slip in and grow the microtubule.

Now, consider a microtubule (Sept and MacKintosh, 2010) that abuts upon a barrier such as a cell membrane whilst polymerizing. Because of thermal motion, occasionally a gap between the barrier and the microtubule will open up and a tubulin subunit can slip in (Fig. 2.9). This increases the compression of the microtubule against the wall. What will the maximum force be?

The tubulin concentration in the microtubule is constant, so the free energy available from polymerization depends only on the free tubulin concentration:

$$\Delta G = \Delta G_0 + RT \ln[\text{tubulin}] \tag{2.8}$$

and the work that must be expended to move the wall is:

$$W = F \cdot d = F \cdot (8\text{ nm/subunit})/(13\text{ protofilaments}). \tag{2.9}$$

So at equilibrium:

$$F = \frac{W}{d} = \frac{(\Delta G_0 + RT \ln[\text{tubulin}])}{0.6\text{ nM}}. \tag{2.10}$$

We are on solid ground so far, but what about under non-equilibrium conditions? For instance, suppose a polymerizing microtubule is moving an object? Can we develop a theoretical force–velocity relation?

One approach is to consider the detailed balance: the chemical and mechanical processes (tubulin association/dissociation, force application through a distance) are separately in equilibrium, and a reaction incorporating both processes is also in equilibrium. The free energy change when a microtubule polymerizes against a force (i.e., mechanical work is performed) will be equal to the sum of the chemical and mechanical free energy change:

$$\Delta G = \Delta G_0 + RT \ln c + F \cdot d, \tag{2.11}$$

where c = [tubulin], and thus

$$\Delta G_{\text{total}} = \Delta G_{\text{chem}} + F \cdot d. \tag{2.12}$$

Thus, for a given value of F, equilibrium is established if $\Delta G_{\text{chem}} = -F \cdot d$, or

$$F \cdot d = -(\Delta G_0 + RT \ln c_{\text{eq}(F)}), \tag{2.13}$$

where $c_{\text{eq}(F)}$ = the tubulin concentration at which polymerization stalls at force, F. Substituting into Eq. (2.12) and rearranging gives

$$\Delta G_{\text{Total}} - \Delta G_{\text{chem}} = -(\Delta G_0 + RT \ln c_{\text{eq}(F)}) \ (= 0 \text{ if } F = 0). \tag{2.14}$$

Since $\Delta G_0 = -RT \ln c_{\text{eq(no force)}}$, the difference between ΔG_{Total} and ΔG_{chem} will be reflected in a change in the equilibrium concentration, c_{eq}, via

$$\Delta G_{\text{Total}} - \Delta G_{\text{chem}} = F \cdot \text{d} = -RT \ln c_{\text{eq}\,(F)} + RT \ln c_{\text{eq (no force)}} = RT \ln \frac{c_{\text{eq(no force)}}}{c_{\text{eq}(F)}}. \tag{2.15}$$

At equilibrium, $k_{\text{on}} c_{\text{eq}} = k_{\text{off}}$ where k_{on} and k_{off} are the on and off rates for tubulin. By substitution,

$$F \cdot d = RT \ln \frac{k_{\text{off(no force)}} / k_{\text{on(no force)}}}{k_{\text{off(force)}} / k_{\text{on(force)}}}$$

$$\frac{k_{\text{off(no force)}} / k_{\text{on(no force)}}}{k_{\text{off(force)}} / k_{\text{on(force)}}} = e^{\frac{Fd}{RT}} \Rightarrow k_{\text{off(no force)}} / k_{\text{on(no force)}} = k_{\text{off(force)}} / k_{\text{on(force)}} \, e^{\frac{Fd}{RT}}$$

$$\Rightarrow k_{\text{off(no force)}} / k_{\text{off(force)}} = k_{\text{on(no force)}} / k_{\text{on(force)}} \, e^{\frac{Fd}{RT}}. \tag{2.16}$$

If we introduce a factor, f, that describes how the extra work ($F \cdot d$) is "split" between the forward and reverse rate constants, the dependence of each rate constant on F can be derived. We define f such that

$$fF \cdot d = RT \ln \frac{k_{\text{off(force)}}}{k_{\text{off(no force)}}} \tag{2.17}$$

then by substitution into equation (2.15):

$$(1 - f)F \cdot d = RT \ln \frac{k_{\text{on(force)}}}{k_{\text{on(no force)}}}. \tag{2.18}$$

So the effect of a force on the polymerization rates will be

$$\Rightarrow k_{\text{off(force)}} = k_{\text{off(no force)}} e^{\frac{fFd}{RT}}$$
$$\text{and } k_{\text{on(force)}} = k_{\text{on(no force)}} e^{\frac{(f-1)Fd}{RT}}. \tag{2.19}$$

One needs to exercise some degree of caution in this kind of analysis since, strictly speaking, thermodynamics does not apply to non-equilibrium situations. In particular, there are two conditions where detailed balance does not apply. The first is if the rates of the reactions in question are not linearly related to the concentrations. For example, if the addition of tubulin subunits were a cooperative behavior, perhaps described by a third-order rate constant:

$$\text{On rate} = k_{\text{on}} \times [\text{tubulin}]^2 \tag{2.20}$$

(Indeed such non-linearity does appear to be the case for microtubules). A second possible weakness is that when the system is out of equilibrium some energy could be dissipated as heat rather than work. For example, under high force a piece of polymer might break off so that it polymerizes/depolymerizes without a load.

2.5 Entropy and Enthalpy

Entropy, S, with units of joules per kelvin is the metric used by engineers to describe the irreversibility of a heat engine. Enthalpy, H, on the other hand, has units of joules, and is the sum of the "internal energy" of a system, U, and the product of the system's pressure and volume. Entropy is defined microscopically as

$$S = -k_{\text{B}} \sum_i p_i \ln p_i, \tag{2.21}$$

where p_i is the probability of microstate state i and the sum is over all such microstates that correspond to the macrostate for which S is being computed. Entropy is defined macroscopically as

$$\text{d}S = \text{d}Q/T, \tag{2.22}$$

where Q is the thermal energy of the system. U and Q are related via

$$\Delta U = Q + W_{\text{mech}} + W_{\text{other}}, \tag{2.23}$$

where W_{mech} and W_{other} represent mechanical and other energies (such as electrical) added to the internal energy of a system. Therefore, thermal energy is one component of internal energy.

Enthalpy, on the other hand is defined as

$$H = U + pV, \tag{2.24}$$

with internal energy defined as

$$\mathrm{d}U = \mathrm{d}Q - \mathrm{d}(pV) = T\mathrm{d}S - p\mathrm{d}V. \tag{2.25}$$

These relationships are valid for continuum mechanics, but have potential pitfalls when applied at the molecular level. Two such examples are the ATP synthase cycle and the kinesin cycle.

As an example, let us assume that ATP-synthase (Fig. 2.8) is operating at 100% efficiency (i.e., the work it performs has the same magnitude as the energy available from ATP hydrolysis). How much heat would it produce? Will it produce any? Our intuition is "no" since all of the energy must be consumed in generating mechanical work (torque). Upon closer inspection however, this conclusion is preposterous. By going through this "proof," we will reveal some of the potential missteps of considering entropy and enthalpy at the level of individual molecules. It must produce heat, since if it were running backward (i.e., pumping protons) it would absorb heat.

The central reaction is

$$\text{ADP} + p_i \rightarrow \text{ATP}. \tag{2.26}$$

The total energy for this at physiological concentrations of ADP, p_i, and ATP may be found using the definition of Gibb's free energy, or $\Delta G = \Delta H - T\Delta S$, and has a value of $\Delta G = 67$ kJ/mol (ATP synthesis requires energy). This equates to about 26 $k_B T$ per ATP molecule.

Now consider how enthalpy changes. From calorimetery measurements (in which the entire enthalpy change is converted to heat), $\Delta H = 48$ kJ/mol, and

$$\mathrm{d}H = \mathrm{d}U + \mathrm{d}(pV) \tag{2.27}$$

$$= \mathrm{d}U + p\delta \mathrm{V} + V\delta \mathrm{p}, \tag{2.28}$$

which represents the change in internal energy plus the work by expansion. However, in a biologic system V and p are usually constant, and this reduces to

$$\mathrm{d}H = \mathrm{d}U \tag{2.29}$$

$$= d\mathrm{q} + d\mathrm{w}$$

or the change in heat plus the work performed. Thus, since at 100% efficiency, $\Delta w = \Delta G = 67$ kJ/mol. This results in a change in heat of $\Delta q = \Delta H - \Delta w = 48 - 67 = -19$ kJ/mol released as heat (or absorbed if ATP is hydrolyzed). Remembering that in this case where $\Delta H = \Delta U$, when ATP is synthesized, 67 kJ/mol must be expended, of which 48 kJ/mol are for the internal energy change (bond formation and bringing together charges when the γ-phosphate associates with ADP) and 19 are to pay an entropic cost (the ATP concentration is relatively high, so the synthesis reaction is entropically unfavorable; recall the thermal forces section) and this results in the simultaneous release of 19 kJ/mol of heat into the surroundings.

This raises the question: Could we use the thermal expansion from this heat to do more work? As it turns out, the answer is "yes," which is puzzling since this apparently violates the second law. So what is going on? Let us imagine a small mechanical system with two states, and an energy difference between these states. For sake of visualization, suppose that the enthalpic energy change is stored as strain (Fig. 2.10a).

The harder we pull on the lever, the more likely a transition to state 2, and if the work expended pulling on the lever equals the energy stored straining the elastic element, the forward and reverse rates will be equal (dashed line in Fig. 2.10c). By this reasoning, it should only take 48 kJ of energy to go between states, so where does 67 kJ (ΔG) come from?

To answer this, we need to consider that the system also depends on the concentration of products and reactants: in a cell there is a relatively high concentration of ATP, which is equivalent to saying our mechanical system is more likely to be in state 2.

For simplicity, suppose what will happen if the system is 10× more likely to be in state 2 (Fig. 2.10b). Now, even though the probability of any given forward transition is equal to the probability of a reverse transition, there are 10× more opportunities to undergo reverse transitions. Thus, even though we are expending enough energy to strain the spring, we are still losing ground. We can restore the balance by applying even more force so that the probability of a given forward transition exceeds the probability of a given reverse transition. But this means we expend more mechanical energy (work) than is stored as strain in the spring, so some of energy driving the forward reaction is lost as heat when the system is at equilibrium.

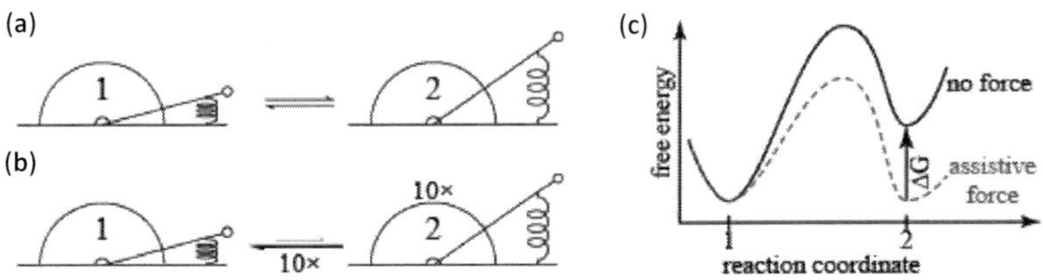

Figure 2.10 Entropic effect on free energy. (a) The enthalpy of the reaction from state 1 to state 2 is stored in the strain of an elastic element. (b) If there are 10 times as many molecules in state 2 than state 1, then there are 10 times as many opportunities for the reverse reaction. (c) The free energy profile of the reaction can be affected by a force assisting the forward reaction. The work this force must do for the forward and reverse reactions to occur at the same rate depends on the enthalpy that is stored in the strain and also on the concentrations of the reactions and products.

But now there appears to be a problem: increasing the force on the lever (so the work expended between states is 67 kJ/mol, the free energy of ATP hydrolysis) also affects the reverse reaction. Specifically, there is only 48 kJ stored as strain, but 67 kJ are required to work against the applied force. Where does the extra energy come from? The answer is of course thermal energy: a (reverse) transition from 2 to 1 will occur if a thermal event provides the required additional 19 kJ of energy.

So why cannot we use the heat released during the forward transition to do additional work? We can...but this reduces the thermal energy available for the reverse reaction! Thus, if we could extract the energy released in the forward reactions, the reverse reactions would still consume heat, and the temperature will drop. The second law is safe!

2.6 Two Ways to Model a Chemomechanical Transition: Macromechanical View versus Statistical Mechanics View—When Do They Apply?

An enormous number of models have been proposed to describe structural transitions in macromolecules, and we will not attempt to summarize them. Instead we present some simple concepts for describing a mechanical transition subject to a load, such as occurs in a motor protein. The intent is to provide the reader with a sense of how the fundamental concepts introduced in this chapter can be applied to describe mechanochemical transitions.

Constant force generator This is the simplest hypothesis, but it can be dismissed since there must be molecular transitions underlying force generation, and with sufficient resolution we should be able to observe these transitions. (Note: despite this, for a short time one investigator promoted their conclusion that kinesin "has no steps" and thus all the current models must be wrong! It turned out that his instrumentation was not as sensitive as he thought it was.)

Constant force stepper Irrespective of the particular mechanism, any motor's mechanotransduction cycle can be divided into two intervals

$$t_c = t_0 + t_\mathrm{d}, \tag{2.30}$$

where t_0 is the portion of the cycle during which no mechanical work occurs, and t_d is the portion where a constant force is produced. For continuous movement down a filament to occur, the motor must move the distance d necessary to reach the next binding site during the interval t_d. The observed speed will be

$$<v> = d/t_c = d/(t_0 + t_\mathrm{d}). \tag{2.31}$$

During t_d the relationship between the force and the speed will be

$$v_{\mathrm{d}} = \frac{F_{\mathrm{m}} - F_{\mathrm{opp}}}{\Gamma} \tag{2.32}$$

where F_{m} is the force during t_{d}, F_{opp} is the applied opposing force (e.g., using optical tweezers), and Γ is the drag coefficient. Since $t_{\mathrm{d}} = d/v_{\mathrm{d}}$, the observed relationship between speed and force will be

$$<v> = \frac{d}{\left(t_0 + \dfrac{d\Gamma}{(F_{\mathrm{m}} - F_{\mathrm{opp}})}\right)} = \left(\frac{t_0}{d} + \frac{\Gamma}{(F_{\mathrm{m}} - F_{\mathrm{opp}})}\right)^{-1} \tag{2.33}$$

The expected relationship between speed and force is linear if $t_0 << t_{\mathrm{d}}$ (i.e., the motor applies force during most of the cycle), convex (down) otherwise.

How could a constant force be generated during the "power-stroke"?

- flexible spring with a large strain ($\kappa x_2 \approx \kappa x_1$)
- multiple state transitions to maintain constant force.

An alternative model is to assume that the motor contains an elastic element of stiffness κ, which, while attached to the filament, is suddenly strained a distance δ (not necessarily equal to the step size between binding sites d) (Fig. 2.11). Directed motion results when this strain is relieved as the motor relaxes. In the absence of an opposing force, the time to complete a move through the distance d can be calculated by applying Newton's second law, neglecting the inertial term so that $\Gamma v = -F_{\mathrm{m}} = -\kappa x$, which can be rearranged and solved:

$$dx = -\frac{F_{\mathrm{m}}}{\Gamma}dt = -\frac{\kappa x}{\Gamma}dt \tag{2.34}$$

$$\Rightarrow \ln\frac{x_0}{x} = \frac{\kappa}{\Gamma}t \quad \text{or,} \quad t = \frac{\Gamma}{\kappa}\ln\frac{x_0}{x} \tag{2.35}$$

The elastic element is applying force from a stretch $x_0 = \delta$ to a stretch $x = \delta - d$, which yields

$$t_{\mathrm{d}} = \frac{\Gamma}{\kappa}\ln\frac{\delta}{(\delta - d)}. \tag{2.36}$$

A constant force opposing relaxation will have the same effect as changing δ by the amount $\Delta\delta = -F_{\text{opp}}/\kappa$

$$t_{\text{d}} = \frac{\Gamma}{\kappa}\ln\frac{\delta - F_{\text{opp}}/\kappa}{(\delta - F_{\text{opp}}/\kappa - d)} = -\frac{\Gamma}{\kappa}\ln\left(1 - \frac{d}{(\delta - F_{\text{opp}}/\kappa)}\right) \tag{2.37}$$

which can be substituted into equations (2.30–2.32) to obtain speed.

The expected relationship between speed and force is slightly convex when unloaded $t_0 << t_{\text{d}}$, and fully convex if $t_0 >> t_{\text{d}}$.

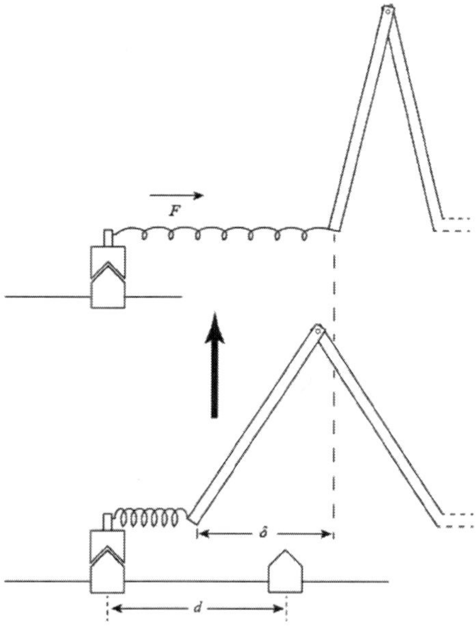

Figure 2.11 A model for a molecular motor walking along a filament. The distance, d, between binding sites on the filament does not necessarily correspond to the distance δ; the elastic element is stretched due to the conformational change in the motor (here depicted as changes in hinge angles).

2.6.1 An Alternative View: Statistical Mechanics

The above models consider the problem of force generation in mechanical terms. An alternative approach is to use statistical mechanics. Consider a state change during which an increment of movement occurs:

$$1 \underset{k_r}{\overset{k_f}{\rightleftharpoons}} 2 \tag{2.38}$$

We begin with the case in which this reaction is essentially irreversible ($k_r << k_f$). The effect of an opposing force on the transition rate can be predicted by the Boltzmann equation:

$$k_f = k_f^{F-0} \exp(-F_{opp}\Delta/k_B T), \tag{2.39}$$

where Δ is the distance from the motor location in state 1 to the position of the potential energy barrier maximum for the transition (see also Fig. 2.6). In terms of the average time of transition this becomes

$$t_d = t_d^{F-0} \exp(F_{opp}\Delta/k_B T). \tag{2.40}$$

Substituting into equations (2.30–2.32) the expected relationship between speed and force is concave if $t_0 << t_d$, convex if $t_0 >> t_d$, and otherwise sigmoidal. If the reaction is reversible then the Boltzmann equation predicts

$$\frac{k_f}{k_r} = K^{F-0} \exp(-F_{opp}d/k_B T), \tag{2.41}$$

where K^{F-0} is the equilibrium constant in the absence of applied force (k_f^{F-0}/k_r^{F-0}). As we discussed previously, to determine the exact effect of a force on the flux through this transition the "split" of the work ($F_{opp}d$) between the forward and reverse rate constants would have to be determined. Nevertheless, it is clear that the average time t_d will be exponentially related to the load, and again the expected relationship between speed and force is concave (up) if $t_0 << t_d$, convex if $t_0 >> t_d$, and otherwise sigmoidal.

All of these models assume that each force generating transition will successfully move the motor by some increment of movement *d*. They make this assumption since continuous movement requires the motor move the increment of movement needed to carry to its next interaction site on a filament. But is this really required? It would also be possible for a motor to fall short of and interaction site, and then make up the difference by diffusion. If this were the case, one would expect that occasionally a motor would diffuse the "wrong way" and end up at the interaction site from whence it started. Under a load the probability of this occurring would increase, and this could slow the motor down.

2.7 Conclusions

Biomolecules experience collisions with other molecules, and the resulting forces have numerous implications, some of which initially seem counterintuitive. Although these thermal forces are complicated and random, we can predict a great deal about their effects. Additionally, taking into account the effects of external mechanical forces and chemical energy inputs yields a range of useful quantitative models for the mechanics of single biomolecules and how they contribute to the physiology of cells and tissues.

Problems

2.1 Perform a literature search on at least three papers to verify the stepping rates of kinesin and myosin. (A) What do you attribute the discrepancies to? (B) Explain any discrepancies between theory and experiment.

2.2 Perform a literature search on at least three papers to verify the energy density of the ATP-myosin system and kinesin. Explain discrepancies.

2.3 Look into the literature to plug values into equations (2.30–2.32) to estimate velocities for other molecular motors.

2.4 Discuss the similarities between Maxwell's Demon and actual organisms.

Suggested further reading: (Berg, 1993; Dill and Bromberg, 2011; Levine, 2009; Oster, 1970).

Chapter 3

Nucleus Mechanics

Kris Noel Dahl[a] **and Markus J. Buehler**[b]

[a]*Carnegie Mellon University, Chemical/Biomedical Engineering, Doherty Hall, 5000 Forbes Avenue, Pittsburgh, PA 15213, USA*
[b]*Massachusetts Institute of Technology, Civil and Environmental Engineering, 77 Massachusetts Avenue, Cambridge, MA 02139, USA*

mbuehler@MIT.EDU

3.1 DNA

3.1.1 DNA Structure

The DNA monomer unit, also known as a nucleotide, consists of a phosphate group, a deoxyribose sugar, and a nitrogenous base. While the sugar essentially contains codon information, it is the phosphate group that provides a local energy source for polymerization during DNA replication. As polymerization is occurring, high-energy molecules, known as deoxyribonucleoside triphosphates (dATP, dTTP, dGPT, dCTP) are hydrolyzed as the polymer forms. Hydrolysis of the PPi from a triphosphate in the monomer to one phosphate per nucleotide in the DNA polymer provides significant energy as nucleotides are added to the polymer. The modification of the phosphate groups also ensures that the DNA polymerization during replication occurs unidirectionally.

Molecular and Cellular Biomechanics
Edited by Bradley E. Layton

ISBN 978-981-4316-83-5 (Hardcover), 978-981-4613-44-6 (eBook)
www.panstanford.com

The resulting polymer is highly negatively charged as a result of the negatively charged phosphate groups between sugars. This large negative charge gives DNA its length-dependent charged character, which is not only useful for laboratory electrophoretic separation, but which also causes self-repulsion and provides unique electrostatic conditions within the nucleus.

However, the structure of the DNA polymer is more complex than a self-repulsive negative polymer. For example, consider the deoxyribose sugar group has a five-carbon ring. Much like the boat-chair transition of six-ring molecules, four carbons are in plane and one carbon resides out of the plane of the molecule (Foloppe et al., 2001). This sugar "pucker" results in transitional states of the DNA in which the phosphate groups on either side of the pucker can be located at either 5.9 Å or 7.0 Å of one another at different equilibrium states (Chary et al., 1987). The 7 Å state is typically entropically and electrostatically preferred. High salt concentration and structural limitations imposed by the nitrogenous base pairing can alter the orientation of the sugar and overall DNA macromolecule as well as the higher-ordered structures of DNA, which we describe in more detail below. Additionally, the unique molecular structures of the four nitrogenous bases, adenine, guanine, cytosine, and thymine, are responsible for the overall DNA structure. The bases of course contain the genetic code for all organisms, as well as structural information for DNA conformation and stability. Pyrimidines (C and T) link via hydrogen bonding with purines (A and G) and pairs create steps within the DNA. AT-connections have two hydrogen bonds, whereas three hydrogen bonds bind the GC pairs. While it is nearly impossible to quantify the change in bonding energy of an individual base pair within a chain, the cooperative effect of several high-energy GC-pairs in a row significantly changes melting temperatures of the two strands. This greater melting temperature corresponds to the greater energy required for the double helix "unzipping," which is performed by the helicase enzyme. Overall DNA shape and histone binding mechanics are also affected by GC content. For example, GC-rich regions force base pairs to stack into a planar structure within the DNA double helix (Dickerson, 1989). To compensate for this stacking, the entire backbone is slightly distorted. Additionally, GC-rich DNA is more likely to form the "A-DNA" structure, which is thicker and has a center hole, compared with "B-DNA," which is the "standard" double helix shape (Wahl and Sundaralingam, 1997). This structural alteration can further affect DNA double-strand flexibility, especially under long-range curvature states.

DNA can be structured as A-DNA, B-DNA, or Z-DNA. Z-DNA is a rare, left-handed form found at high salt concentrations, which alter the electrostatic repulsion (Leng, 1985). A- and B-DNA, the predominant forms, are right-handed helices that are 2.5 nm in diameter with 11 bp/turn. Major and minor grooves are formed from the double strands of DNA and allow for structural heterogeneity for protein binding along the genome. The double helix also provides for redundancy in the genetic code, allowing for proofreading during replication and repair against DNA damage since point mutations have incorrect pairing which leads to DNA destabilization (Kunkel and Erie, 2005). Unlike proteins, which rely on DNA to be properly made, in order for an organism to remain robust, it is critical that the genetic code remain unaltered despite insult from chemicals, radiation, etc. As such, having multiple mechanisms to maintain DNA's fidelity is important for host survival and propagation.

3.1.2 DNA Mechanics

The one-dimensional mechanical measurements of DNA fall into different regimes based on the level of extensibility. For longer length scales, a majority of the DNA mechanical properties are defined by removal of entropic bends or undulations (i.e., pulling the DNA straight when it wants to explore a bent (more entropic) conformation). The length scale over which a molecule such as DNA is rigid before it bends is called the "persistence length,"

$$\lambda = \frac{EI}{k_{\mathrm{B}}T}, \tag{3.1}$$

where E is the elastic modulus, I is the area moment of inertia and $k_{\mathrm{B}}T$ is thermal energy (k_{B} is Boltzmann's constant and T is temperature). The expression of equation (3.1) may be viewed as the ratio between the product of a molecular structure's capacity to sustain strain energy, E and its girth, I normalized by its natural propensity to explore a multitude of states, $k_{\mathrm{B}}T$. Note that the product of E and I is also commonly referred to as the bending modulus, $\kappa = EI$. For DNA, $\lambda = 53 \pm 2$ nm, which has been measured with many imaging modalities and physical manipulation (Hegner and Grange, 2002). Two biological factors related to the nucleosome fall out of the biophysics of DNA's persistence length. While 50 nm is a short persistence length compared to other biological filaments, it is large compared to the length scales upon which DNA interacts within the nucleus. For example, DNA winds around histones, which are 10 nm in diameter. It is energetically unfavorable for DNA to mechanically bend around this 10 nm contour. However, the histone's significant positive charge attracts the negatively charged DNA, thereby overcoming DNA's resistance to bending. This interplay of mechanics and charge allows for unwrapping of DNA from histones when charges are locally screened, such as by complexes which remove DNA from histones during transcription (Montecino et al., 2007).

At high strains or short length scales, DNA can also undergo other deformations such as extension, torsion, and torsion-stretch coupled extension. These deformation modalities are only significant after the DNA has been pre-strained to remove any entropic bends. However, these deformations are extremely important in calculating forces required for functional DNA processing which require molecular motors to produce extension and twist on DNA at the length scale of a few base pairs. DNA extension has been measured by microneedles, magnetic pull, fluid flow, and optical traps (Bustamante et al., 2000). Initial stretching of long DNA molecules measures entropic straightening of the molecule related to κ. At these small deformations, DNA straightens with much less than 0.1 pN of force in a completely reversible way (Smith et al., 1992). At medium extensions (1 pN < force < 65 pN) DNA extends reversibly, even slightly beyond its predicted contour length. Thus, the unique structure of the double helix structure permits additional extensibility to the polymer. The stretch modulus (force/stretch) for this medium regime is greater than 1000 pN (Smith et al., 1996; Wang et al., 1997). Beyond 65 pN of applied force, DNA becomes "overstretched" and exhibits a hysteresis when relaxed, probably a function of lost double helix structure and related base-pair uncoupling (Storm and Nelson, 2003). Torsion and torsion-stretch have also been measured for the medium, reversible regime. To compare with the persistence lengths, moduli are normalized to k_BT to produce effective lengths. Torsion length is 120 nm and torsion-stretch length is much lower at 50 nm (Moroz and Nelson, 1997). The relatively soft torsion-stretch modulus, similar to bending, shows the benefit of combining twist with extension for a helical molecule.

3.1.3 DNA Assembly into Chromatin

DNA rarely exists as a single-double helix for any significant length scale in vivo. DNA winds around histones to package the nearly 1 m of DNA into the nucleus and allows modifiable access to regions of DNA for expression. The DNA–histone complex, called the nucleosome, contains a 10 nm histone octamer (made of core histone proteins H3, H4, H2A, and H2B) which is wound two-times by DNA (Richmond et al., 1984). The DNA sequences that bind to histones statistically show multiples of GC and (AA, TT or TA) (Anderson and Widom, 2001; Kaplan et al., 2009). This may be related to the mechanical character of bending that GC-content allows, but there is increasing evidence from bioinformatics that the DNA sequence is related to the probability of transcription (Segal et al., 2006).

To quantify the mechanics associated with DNA wrapping onto histones, experiments have examined both association and dissociation. To measure association, DNA strands were exposed to flow fields in a solution that contained histone octamers. The flow field-induced force can be compared to the speed of the resulting decrease in DNA length as it coils. Thus, it is possible to approximate the force of DNA-association onto a nucleosome. Conversely, pulling on chromatin with an AFM or laser tweezers shows stochastic jumps associated with nucleosomes popping off of the chromatin structures.

The interlinked nucleosomes order into a thicker fiber called chromatin. Isolated and reconstituted chromatin has the appearance of 10 nm-thick "beads on a string" in its extended conformation. When condensed, chromatin forms a 30 nm-thick fiber (Ostashevsky, 2002). The linker histone H1 plays a significant role in this condensation (Trojer and Reinberg, 2007). Within the nucleus chromatin is segregated structurally and functionally into euchromatin and heterochromatin. Euchromatin is more easily translatable and is transcribed first (Gilbert et al., 2004). Structurally, euchromatin is open and nuclei with mostly euchromatin, such as in stem cells, are very compliant (Pajerowski et al., 2007). Conversely, heterochromatin is in a condensed state. Genes in heterochromatin are rarely expressed, and DNA in heterochromatin is replicated late in S-phase. By virtue of the fact that it is a dense structure, heterochromatin is stiff. Within the nucleus, heterochromatin is a load bearing structure and sudden ablation can cause nuclear collapse (Mazumder and Shivashankar, 2007). These mechanical changes are related to the expression of heterochromatin-specific proteins, including isoforms of Heterochromatin Protein 1 (HP1), and modifications of DNA and histones. DNA and histones in the heterochromatin are modified by acetylation, methylation, and phosphorylation. For example, trimethylation of histone H3 lysine 9 is commonly known to be associated with heterochromatin (Georgatos et al., 2009). In general, chromatin found in heterochromatin possess a higher density of nucleosomes per DNA than in euchromatin which is closer to the 30-nm fiber seen in vitro (Bassett et al., 2009). While regions of compact chromatin can occasionally contain active genes and open chromatin can contain silenced genes, statistically this is not the norm (Gilbert et al., 2004).

At the nuclear periphery, heterochromatin structures are bound to and are stabilized by proteins of the nucleoskeleton called lamin proteins (Gruenbaum et al., 2005). Lamins can interact with chromatin directly via nonspecific 30–40 bp DNA segments (Moir et al., 2000a; Stierle et al., 2003) and by binding to histones or via lamin-binding proteins (Segura-Totten et al., 2002). A specific class of inner-nuclear membrane-lamin-binding proteins called LEM-domain proteins binds DNA through BAF (Barrier to Autointegration Factor). Another lamin associated protein, known as the LBR (lamin B-receptor), binds to lamin B, DNA, histones, and HP1 (Ye et al., 1997; Worman, 2005). Lamin structures inside the nucleus may also serve to stabilize heterochromatin throughout the nucleoplasm. Chromatin is also organized into larger chromosome structures, but chromosomes themselves are not generally considered as mechanical units within the interphase nucleus.

3.1.4 Mechanics of the Nuclear Interior

The combination of heterochromatin and euchromatin at the nuclear interior produces a mixture which can be modeled as a heterogeneous viscoelastic material (Dahl et al., 2005b; Dahl and Kalinowski, 2011). Unlike traditionally studied rheological materials, which behave as continua, the nuclear interior has internal movements directed by discrete molecular motors such as helicases, polymerases, and myosin. Furthermore, the organization of individual genes in the nucleus is non-random (Parada et al., 2004), lending further contrast to most self-assembled soft materials, even ones with higher-order structures.

To sufficiently model the mechanics of the genome at this level, some investigators have used traceable injected markers within the nucleus to decipher long-scale reorganization. These studies reveal a viscosity and elasticity of the intranuclear space of 3T3 fibroblasts to be 520 poise and 180 dyne/cm^2 (18 Pa), respectively (Tseng et al., 2004). These observed properties are much greater than the viscoelastic values measured of the cytoplasm, which are more similar to values obtained in studies of whole nuclei within living cells (Guilak et al., 2000b; Caille et al., 2002b; Dahl et al., 2005b). Further work using injection of traceable markers into the nucleus is warranted since it has the ability to further reveal the heterogeneous, organized interior nucleus with its many motor proteins interspersed within regions of low viscosity induced by the injected fluid.

Most studies aimed at measuring the contribution of the genome to nuclear mechanics typically require large applied forces and deformations. Micropipette aspiration, which is used to deform either isolated nuclei or nuclei in cells with chemically disrupted cytoskeletons, has shown that the nucleus is extremely stiff. By altering DNA density or content in nuclei to differing extents, micropipette aspiration has shown that the chromatin contributes to the viscoelastic behavior whereas the lamina, discussed below, provides elastic resistance (Dahl et al., 2004a). Using micropipette aspiration with photobleaching of fluorescence chromatin markers, chromatin flow can be observed under tension (Pajerowski et al., 2007). Recent studies of laser-ablation of heterochromatin structures within the nucleus have shown that heterochromatin is a pre-stressed load bearing structure necessary for the integrity of the nucleus and the cell (Mazumder and Shivashankar, 2007). Atomic force microscopy (AFM) and other biophysical techniques have been used to study nuclear mechanics, but it still remains difficult to decouple the rheological properties of the nuclear interior from the mechanics of the nucleoskeleton and lamina (see Section 3.3.1 for a discussion of whole-nucleus measurements).

3.2 Lamins

3.2.1 Lamin Protein Structure

The genomic DNA of the nucleus is protected and regulated by the nucleoskeleton, which lies primarily at the inner nuclear membrane. The nucleoskeleton has many classes of elements including spectrin-like elastic filaments, a nuclear isoform of titin and pore-linked filaments. The human nucleoskeleton is primarily composed of the lamins, which are type V intermediate filaments. There are two classifications of human lamins: A-type and B-type. The A-type include primarily lamin A and lamin C. These are splice variants of the gene *LMNA* and are present at various levels in a wide range of tissues. B-type lamins include lamin B1 and B2, which are produced from different genes *LMNB1* and *LMNB2* (Bridger et al., 2007) and which are required for cell and animal survival (Harborth et al., 2001; Vergnes et al., 2004). Loss of *LMNA* is not lethal, but is associated with severe muscular dystrophy (Harborth et al., 2001; Raharjo et al., 2001). Interestingly, the loss of lamin A but not lamin C (the lamin C-only phenotype or $lmna^{LCO/LCO}$) in mouse model systems shows no overt pathology (Fong et al., 2006).

Similar to other intermediate filaments, lamin proteins have a globular N-terminal head, a central α-helical rod region and a globular C-terminal tail. The N-terminal head domain consists of only 31 amino acids, and shows low conservation between A-type and B-type lamins. The rod domain (354 amino acids) is highly conserved among lamins. The singular stable unit of the lamin is a dimer, which is stabilized by coiled interactions of the rod domains. The largest difference between lamins and other intermediate filaments is in the rod domain, which can vary by 278 amino acids or more and is 40% larger than other intermediate filaments (Herrmann et al., 2003; Herrmann et al., 2007). Lamins A, B, and C also show variation in the length of the tail domain. The large tail domain is mostly disordered except for an s-type Ig fold (Dhe-Paganon et al., 2002; Krimm et al., 2002). This disorder may aid in allowing binding of lamins to a variety of binding partners (Zastrow et al., 2004a).

Lamin tail domains also undergo a complex post-translational modification, likely to aid in nuclear envelope localization (Young et al., 2006). The tail region of lamin A, lamin B1 and lamin B2 is farnesylated and carboxymethylated by a ubiquitous cellular prenylation process (Young et al., 2005b). Lamin C, the smaller protein product from *LMNA*, does not undergo such processing. Lamin A is subsequently processed to remove the lipidated C-terminus by an endoprotease Zmpste (Young et al., 2005b). Alterations of the processing of lamin A can lead to altered nuclear structure and, in some cases, severe diseases (Young et al., 2005b). If lamin A cannot be processed, intranuclear aggregates can accumulate inside the cell. The retention of the lipidation on prelamin A (lamin A prior to Zmpste-cleavage) can be caused by a mutation in *LMNA* or in the Zmpste gene. This in turn leads to premature aging or restrictive dermopathy, respectively (Young et al., 2005a).

3.2.2 Filament Assembly and Network Properties

As with other intermediate filaments, lamin assembly from protein to dimer to mature filament is a complex process. In vitro polymerization into the mature filament and filament network has not yet been achieved for human or mammalian lamins, which has thus limited the study of their mechanical properties. Lamin dimerization has been visualized, as has the heat-to-tail assembly; the lamin head-to-tail assembly is different than the typical head-to-head and tail-to-tail assembly of other intermediate filaments (Stuurman et al., 1998). Charge clusters on lamins may be responsible for their unique assembly structure (Strelkov et al., 2004). Generally, charge plays important roles in lamin assembly and disassembly: phosphorylation of lamins leads to disassembly during mitosis and nuclear envelope break down (NEBD). From these assembled dimer chains, intermediate filaments typically form staggered lateral associations making a rolled-up sheet of an intermediate filament (Buehler and Ackbarow, 2008; Goldman et al., 2008). Unfortunately, lamins form ordered paracrystalline arrays in vitro, which show no in vivo counterpart (Moir et al., 1991; Heitlinger et al., 1992).

The only observed organized lamina network occurs in amphibian oocytes as a two-dimensional, orthogonal network (Aebi et al., 1986; Akey, 1989). Electron micrographs of *Xenopus* oocyte lamina showed a meshwork with a thickness of 10-20 nm. The lamin filaments had a diameter of 10.5 ± 1.5 nm and a length of 50 nm (Aebi et al., 1986). These specialized lamina networks have also been used as template for assembly of human A- and B-type lamins. Lamins were injected into the *Xenopus* oocyte nucleus, and filaments were then visualized using electron microscopy. A-type lamins in this system form thick sheets and bundles, and B-type lamins assemble into more regular, dispersed structures (Goldberg et al., 2008). Using this template system, lamin filament diameters were approximated to be 10 nm for lamin L_{III} (the endogenous *Xenopus* oocyte lamin), 7 nm for lamin B2, and 15 nm for lamin A (Goldberg et al., 2008).

In addition to assembly differences, there are other structural and organizational differences between A- and B-type lamins (Goldberg et al., 2008). As discussed above, A- and B-type lamins are present at different levels within vertebrate cells and are differentially important for cell and organism survival. After cell division, B-type lamins are incorporated into the nucleoskeleton at a different time than A-type lamins (Moir et al., 2000b). Wild-type lamins A and B1 have also been observed via time-domain fluorescence lifetime microscopy to polymerize in distinct homopolymers that further interact in the nucleoskeleton (Delbarre et al., 2006). During interphase, A- and B-type lamins show different dynamics as shown through FRAP experiments to GFP-tagged lamins expressed in mammalian cells (Daigle et al., 2001; Dahl et al., 2006). Forster resonance energy transfer (FRET) studies also suggest that lamin filaments exist as homopolymers of the same protein in normal cells (Delbarre et al., 2006). Generally, the unique spatial and temporal organization of lamin A and lamin B suggest minimal interaction despite high levels of homology and close association.

3.2.3 Other Nucleoskeletal Proteins

In addition to forming filaments that lead to network structures, lamins bind to other proteins. Lamin associated proteins (LAPs) stably bind with lamins and can be co-extracted with lamins from nuclei treated with detergents and salts (Senior and Gerace, 1988; Foisner and Gerace, 1993). LAPs can be located at the inner nuclear membrane or in the chromatin nucleoplasm (Dorner et al., 2007). LAPs in or at the inner nuclear membrane associate with the lamina nucleoskeleton and support structure and integration of structures. Prelamin A associates with the nuclear membrane via farnesylation, but LAPs are required for membrane association after cleavage of this tail. There are more than 80 putative nuclear envelope proteins have been identified to directly or indirectly interact with lamins (Schirmer and Gerace, 2005). Well-studied LAPs include LAP1, LBR (lamin B-receptor), SUN1, SUN2, nesprins, and LEM domain proteins (Holaska et al., 2002; Zastrow et al., 2004b). LEM-domain proteins (with a conserved 40 amino acid sequence) include LAP2 (now known as LAP2β), emerin, MAN1, and LEM2/NET25 (Wagner and Krohne, 2007; Huber et al., 2009). Only LAP1C, LAP2β, and LBR preferentially interact with B-type lamins (Gotzmann and Foisner, 1999). The other proteins either primarily or exclusively bind A-type lamins. The multiple binding partners of A-type lamins may suggest changing binding partners based on signaling.

3.2.4 Actin

Actin is present within the nucleus, but in contrast to that found in myocytes, nuclear actin is likely not polymerized beyond a few monomers (Gieni and Hendzel, 2009). Actin filaments found in the nucleus are suggested to form from unique polymerization type that cannot be detected by phalloidin (Gieni and Hendzel, 2009). From every investigated avenue, it appears that actin is not acting as a structural filament in the nucleoskeleton, but it could be reinforcing lamin structures or acting as a scaffolding element. Also, nuclear actin is involved in mRNA synthesis, processing, and transport and chromatin remodeling, often in conjunction with nuclear myosins (Pederson and Aebi, 2002; de Lanerolle et al., 2005).

3.2.5 Nucleoskeleton Mechanics

The lamina nucleoskeleton is a stiff, load-bearing element necessary for the structural integrity of the nucleus. Measurement of

Xenopus oocyte nuclei, which have a low density of DNA structures, has a two-dimensional lamin elastic modulus of the lamina of 25 mN/m. However, this characterization may reflect the "perfect" organization of the L_{III} lamins inside the oocyte, described above. Associated to the lamina, chromatin also participates in defining the mechanical properties of the nucleus, described in Section 3.3. However, after osmotic swelling isolated nuclei, it has been observed that the nuclear lamina swells independently of the underlying chromatin (Dahl et al., 2004b; Dahl et al., 2005c).

Intermediate filaments generally show a more viscoelastic deformation under stress than other cytoskeletal filaments (Janmey et al., 1991). This is likely due to the rope-like, staggered assembly compared with the linear polymerization of actin and microtubules. The intermediate filament molecular structure also shows transitions from α-helix to β-sheet along the length of the rod domains (Qin et al., 2009; Qin et al., 2010). This transition of the rod domain leads to a thinning and stiffening of the dimers within the filaments and network. Network modeling of these strain-induced transitions have recently been performed, and suggest that they account for mechanical deformation and flaw tolerance associated with the nucleoskeleton (Qin and aBuehler, 2011).

Many studies have shown that lamin A is primarily responsible for influencing nuclear mechanics, compared with lamin B. Functional loss of lamin B (lamin B1), while impacting survival, does not impact nuclear mechanics (Lammerding et al., 2006). Removal of A-type lamins or their analogs result in weak nuclei. Lamin-depleted nuclei isolated from *Xenopus* egg extracts are fragile (Newport et al., 1990), and nuclei from murine *lmna*$^{-/-}$ cells rupture with applied stress (Lammerding et al., 2004). Extra exogenous expression of lamin B1 is also unable to restore the nuclear mechanical properties of *lmna*$^{-/-}$ mouse embryo fibroblasts (Stewart et al., 2007). Leakage of chromatin from the nucleus into the cytoplasm is often observed within the *lmna*$^{-/-}$ cells. Global cell mechanics is also altered in lamin A-reduced cells (Broers et al., 2004a; Lammerding et al., 2004; Lammerding and Lee, 2005). This is likely from the loss of the nucleoskeleton-cytoskeleton connections (Broers et al., 2004a). Thus, the nucleus may act to organize and anchor the actin cytoskeleton at the center of the cell.

Lamin C is structurally and mechanically different than lamin A despite their genetic similarity. Lamin C shows higher localization in the nucleoplasm versus the nucleoskeleton at the inner nuclear membrane (Lammerding et al., 2006). The mechanical properties and morphology of nuclei in *lmna*$^{LCO/LCO}$ fibroblasts were only slightly different than nuclei of *lmna*$^{+/+}$cells, especially compared to *lmna*$^{-/-}$ nuclei (Fong et al., 2006). However, the mice were in every other way normal and did not present the muscular dystrophy phenotype seen in *lmna*$^{-/-}$ mice. This result suggests that lamin C may be enough for the general function of the nucleus and may propose a therapy for some laminopathies as an alternative to the use of farnesyltransferase inhibitors, where prelamin A and lamin A may be removed through RNAi technology or oligo gene therapy (Scaffidi and Misteli, 2005). However, it is unclear what implications that this small change in nuclear mechanics may have long-term. The mechanisms underlying the organization of the lamina are poorly understood and require more information about the physical properties of its single components and of its interactions in vivo.

Useful insight into nuclear structure and mechanics can also be gained from situations where there is an effective increase of lamins, which leads to an increased stiffness of the nucleus. Mutations in the *LMNA* gene related to an abnormal post-translationally modified lamin A, described above, cause accumulation of lamins at the inner nuclear membrane (Goldman et al., 2008). The over-accumulation and slow turnover of mutant lamin A at the membrane also entrains wild-type lamin A (Dahl et al., 2006). This effective increase in lamins at the nuclear envelope results in a stiffer lamina network, which is less able to deform under tension (applied by micropipette) or recover after osmotic swelling. Thus, there is an optimal level of lamins in a cell, and excess mechanical stiffness can be just as deleterious as loss of mechanical strength. The lamin nucleoskeleton protects the nuclear interior, but also facilitates deformation and adaptation.

Altered lamin concentrations also lead to changes in nuclei shape, which can be used as a diagnostic tool (Lammerding et al., 2007a; Choi et al., 2011). Absence of lamin A results in altered nuclear shape, decondensation, and homogenization of chromatin, as well as altered distribution of LAPS and B-type lamins (Sullivan et al., 1999; Raharjo et al., 2001; Nikolova et al., 2004). Lamin-chromatin binding also plays a role in maintaining structure and mechanical integrity of the cell nucleus. This suggests that both nuclear regions participate cooperatively in defining the mechanical properties of the nucleus, where the contribution of one region versus the other may vary depending on nuclear structure, morphology and the surrounding mechanical environment.

3.3 Whole Nucleus Properties

3.3.1 Whole Nucleus Mechanics

The nucleus is the largest and stiffest organelle in the cell (Dahl et al., 2010). Unlike the actin cytoskeleton, the nucleus is not capable of rapid reorganization. Thus, the mechanics of the cell often restrict cellular motility and movement. Also, the nucleus appears to act as a central mechanical element that aids in cytoskeletal force generation from one side of the cell to the other (Chancellor et al., 2010). Interactions of the centrosome and nucleus can also impact polarity and directionality of cellular transmigration (Chancellor et al., 2010; Luxton et al., 2010).

Whole nucleus mechanics can be explored in numerous ways, and all have limitations. Some experimental methodologies examine isolated nuclei (Guilak et al., 2000a; Caille et al., 2002a; Dahl et al., 2005a). This typically allows for more direct stress–strain measurements at the expense of modifications caused by an altered chemical environment. Other methods include measurements on cells with depolymerized cytoskeletons (Rowat et al., 2005; Ribeiro et al., 2012). However, the mechanical integration of the cytoskeleton and nucleoskeleton may cause alterations to nuclear structure and mechanics. Lastly, the entire cell can be examined with visualization on the nucleus. This is limited by dispersion of forces through the cellular cytoskeleton and a poorly defined strain field (Lammerding et al., 2007b).

Micropipette aspiration measures the mechanics of isolated nuclei and nuclei in cells with depolymerized cytoskeletal structures (Dahl et al., 2005b; Rowat, 2009). Since the length scale examined is on the order of the whole nucleus, the nucleus can be considered a relatively course-grained homogeneous material. Micropipette aspiration also allows for visualization of subnuclear domains during the mechanical measurement. Uniaxial compression to investigate cellular and nuclear deformations under load was first conducted in chondrocytes in articular cartilage (Guilak, 1995; Knight et al., 2002). Compression experiments on single cells and isolated nuclei, using a microindentor or microplate, have been used to study cellular and nuclear biomechanics in more detail (Caille et al., 2002b; Broers et al., 2004b). Atomic force microscopy (AFM) has also been used on isolated nuclei and cells, but with mixed results (Dahl et al., 2005b; Moreno-Flores et al., 2010).

Nuclei exhibit a viscoelastic response to applied force. Some models suggest nuclei are purely elastic (Deguchi et al., 2005), while others involve three-parameter spring-dashpot models (Guilak et al., 2000b), or power-law rheological models that are equivalent to an infinite series of Kelvin solids (Dahl et al., 2005b). Generally it is not clear how the different mechanical elements of the nucleus are mechanically interconnected with one another and with the cytoskeleton. Often, biological perturbation of one structure leads to structural and functional changes downstream, which limits singular investigation of structures. Typically, comparative investigations between normal and abnormal cells are considered. However, there is as yet no standard for measuring or quantifying nuclear mechanics.

Problems

3.1 How can DNA, with λ = 53 nm, wrap multiple times around a histone which is 10 nm in diameter?

3.2 If the radius of gyration is the effective sphere that a polymer sweeps out from entropic fluctuations. For a semi-flexible polymer the radius of gyration is a function of the persistence length (λ) and the contour length (L_c, total length) by $R_g = (2\lambda L_c)^{1/2}$. How much space would the genome (1 m) take up if it were unmodified? Compare this with a nucleus with a diameter of 10 μm.

3.3 The nucleoskeleton, a network of intermediate filament proteins, can be modeled simply as a three-component viscoelastic material. Consider a three-component viscoelastic model with a spring (A) in parallel with a spring (B1) and dashpot (B2) in series.

(a) Draw the system

(b) Would this model be considered a viscoelastic solid or viscoelastic fluid? Why?

3.4 We perform a creep test on the nucleoskeleton to determine its mechanical properties. One advantage to a testing a cellular system is that we can alter the biology by removing proteins (knockdown with siRNA) or add in extra proteins (with expression of exogenous recombinant DNA). Consider a creep test on the 3 component viscoelastic model described above.

(a) Graph the stress versus time

(b) We increase the levels of one of the nucleoskeletal structures with the expression of GFP-lamin A. Graph the strain versus time if the lone spring (A) is significantly stiffer than the B1–B2 combination.

(c) We remove the protein with the expression of an siRNA for the LMNA gene, which codes for lamin A. Graph the strain versus time if the spring and dashpot (B1–B2) are significantly stiffer than the lone spring (A).

Chapter 4

Nanoscale Imaging and Modeling

Bradley E. Layton,[a] **M. Brent Boyd,**[b] **and Kathleen B. Allen**[b]

[a]*Applied Computing and Electronics Department and Biochemistry and Biophysics Program, The University of Montana, 32 Campus Drive, Missoula, MT 59812, USA*
[b]*Mechanical Engineering and Applied Mechanics Department, Drexel University, 3141 Chestnut St, Philadelphia, PA 19104 USA*

bradley.layton@alum.mit.edu

As we saw in the previous chapter, organization of the genome into the information-dense, yet accessible DNA molecule is critical to life, for which in Chapter 1 we used the working definition of "an entropy partitioning system." This chapter examines some of the structures that cells use to spatially partition their internal structures such as tubulin, structures that cells uses to maintain their boundaries such as the cell membrane, as well as structures that allow cells to maintain their spatial distribution among other cells such as collagen.

4.1 The Structures of Entropy Partitioning

In this chapter, we use atomic force microscopy as lens through which we can study nanoscopic structures such as collagen

Molecular and Cellular Biomechanics
Edited by Bradley E. Layton

ISBN 978-981-4316-83-5 (Hardcover), 978-981-4613-44-6 (eBook)
www.panstanford.com

fibrils, microtubules, and liposomes. These three were selected as model systems because they may be regarded as quasi-static mechanical structures that are responsible for maintaining spatial relationships among cells and subcellular components. All three of them are produced by metabolic processes and are thus relied upon by living systems to maintain a spatial organization among individual cells (collagen), material gradients (polymerized tubulin), or entropy and concentration gradients (cell membranes). Thus, the information inherent in the spatial position and orientation of all of the molecular subunits (amino acids and phospholipids) of each of these three structures is inherent to allowing a living system to maintain a large entropy gradient, ΔS via relatively little information content, I, implying a large α, which as we argued in Section 1.4.1 is preferable for survival.

Of particular interest is the relationship between the tertiary structure of individual proteins and their mechanical properties.

4.1.1 Collagen

Collagen, the most ubiquitous protein in mammals (Ricard-Blum, 2011), appears to have been "searching" genomic space looking for a structure that is capable of binding to itself in either a triple helix with cleaved N and C termini, or in afibrillar structures. With at least 38 genes in the human genome (Myllyharju and Kivirikko, 2001; Baronas-Lowell et al., 2003), and its prevalence on a global mass basis, its central role in maintaining multicellular life cannot be understated. More recently, its role in maintaining the structural organization of "lower" organisms such as hydra (Dansky and Johnson, 1986; Shimizu et al., 2008) and cyanobacteria (Layton et al., 2008; Price, 2013; Price and Anandan, 2013) as well as other prokaryotes (Caswell et al., 2008) has been of interest.

The precise singular path (if in fact there is one) by which these prokaryotes came to contain and express collagens is not known, and the pathways by which their expression is regulated is far from well understood. However, one could imagine constructing a cost–benefit function for the expression of collagen (or any other protein for that matter), using the theories of Darwin and Dawkins as guiding principles (Zeiger and Layton, 2008). Some of collagen's roles are highlighted in Table 4.1. For a more exhaustive list of collagen's multifaceted roles, see Layton (2003a).

Table 4.1 A few key mechanical/structural proteins and their locations and roles within cells

Protein	Location	Tissue	Role
Actin	cytoskeleton	all cells except nematode sperm	compression bearing tension bearing
Collagen type I	extracellular	bone, ligament, cornea, nerve, skin	maintenance of tensile force; maintenance of compressive force if calcified
Collagen type II	extracellular	cartilage	maintenance of compressive force
Elastin	extracellular	arteries, lungs, skin, bladder, cartilage	long-range force interactions
Lamin	nucleus	all	dna binding; nuclear pore interactions
Spectrin	sub-cellular membrane	erythrocytes	shear bearing
Tubulin	nucleus, axons	all, esp neurons	compression bearing, transport

To maintain multicellularity, an extracellular matrix (ECM) structure such as cellulose and collagen (Fig. 4.1) is necessary. Without an ECM, whose primary role is to maintain spatial organization among cells, other physical pathways, including chemical, electrical, and thermal networks among and between cells of a given organism or colony, would not be maintained and the organism would quickly disintegrate.

As we have seen in the preceding chapter, there are also a great number of roles for structural mechanics within individual cells and indeed within individual nuclei. Revealing the molecular events of cell growth is critical to obtaining a deeper understanding of tissue and organ development, injury response, and for tissue engineering applications. Central to this is the need to understand the mechanical interactions between the cytoskeleton and the cell membrane and how these interactions affect the overall growth mechanics of cells. We will visit this topic in detail in Chapter 5.

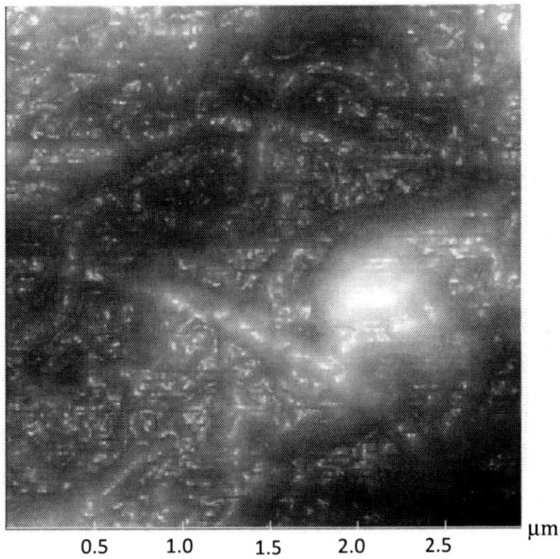

Figure 4.1 Atomic force microscopy image of collagen fibril-like structures from *Trichodesmium erythraeum*. We have hypothesized that this unique collagen-like molecule is responsible for maintaining colony structures as large as several square kilometers (image taken by Brent Boyd. TE courtesy of Simara Price).

4.1.2 Cellular "Safety Factor"

We recently proposed that a "natural mechanical safety factor" exists that allows for the cytoskeleton to impose forces upon the cell membrane without causing cell damage (Allen et al., 2009a). In the same way that a garbage bag might be engineered to hold its various contents without tearing, the cell membrane must also be simultaneously tough (strain energy per volume prior to failure) and resilient (strain energy per volume prior to elastic limit). Forces generated between the cytoskeleton and the cell membranes are governed by the respective mechanical properties of both, and a balance between these forces is necessary for the cell to grow, divide, move, etc. If the cell membrane were inflexible, cell motility would be impeded. However, if it were too flexible, the membrane would be likely be commensurately weak, and it would be unable to sustain environmental forces, such as shear stresses due to fluid flow, as well as internal forces such as cytoskeletal movements. Similarly, if the cytoskeleton were too strong, interaction with the cell membrane might result in membrane rupture and cell death. However, if the cytoskeleton were too weak, it would not be able to exert enough force on the membrane to sustain cell shape or promote cell growth (See Q4.1 at the end of this chapter for discussion on stiffness versus strength).

4.1.3 Tubulin and Actin

The peripheral nervous system demonstrates the importance of a mechanical balance between the cytoskeleton and the cell membrane. Peripheral nerve cells contain extraordinarily high-aspect-ratio axons that may extend for meters, with diameters six orders of magnitude smaller (Dennerll et al., 1988; Ingber, 1993; Baas, 2000; Tuszynski et al., 2005). In order for these structures to exist, axons contain microtubules, actin filaments and intermediate filaments that work together to permit axons to grow, and subsequently to maintain their shape once growth is complete. However, rather than existing in a static state, these proteins are composed of living structures that are genetically "programmed" for self-assembly via mobile protein subunits, and these protein struts generate mechanical forces between and among one another (Hotani and Miyamoto, 1990). The dynamic equilibrium of forces on the cytoskeletal elements integrate the cytoskeleton into a structure with mechanical integrity and stability as well as mechanical and metabolic plasticity to respond to biological cues such as stretch or injury. It is the interactions between the actin filaments and the neuronal membrane that are responsible for mechanically moving the growth cone forward during development (Andersen and Bi, 2000). The cell's mechanical properties (e.g., stiffness, resting tension, etc.) also result from the interactions of these structural proteins (Ingber, 2003b).

The remarkable structure of individual microtubules has been revealed by atomic force microscopy (Figs. 4.2 and 4.3) (Vater et al., 1995).

In their study on microtubule assembly and interaction kinetics, Hamon et al. (2010) viewed tubulin protofilaments as well as microtubule dispersions on mica.

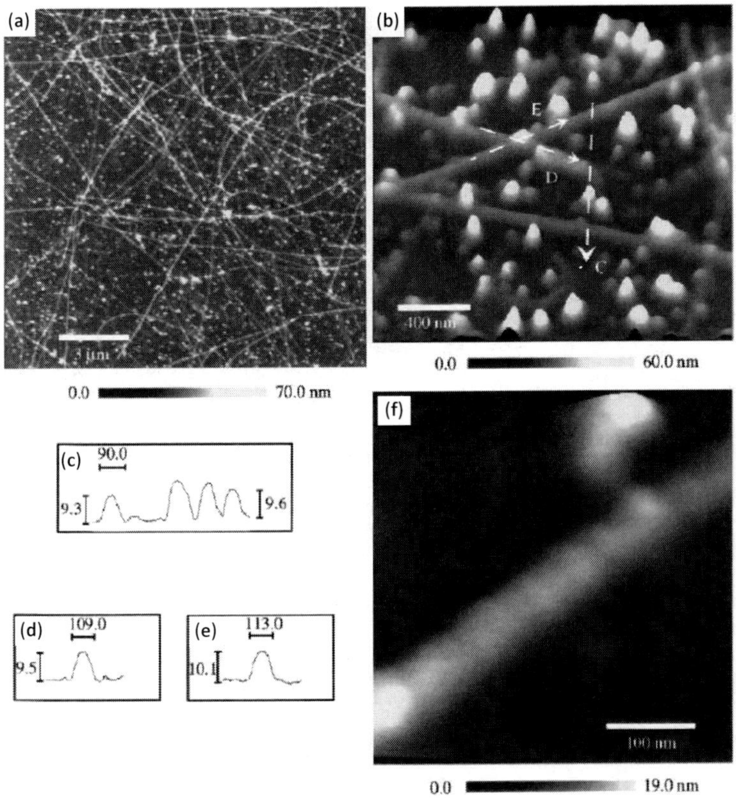

Figure 4.2 Air contact mode atomic force microscopy image of microtubules derived from porcine brain tissue on a polylysine surface. In (a) individual MTs can be seen spanning the entire 18 mm field of view. (b) a zoomed-in 2 µm field of view titled by 75° of a small subset of individual MTs with the cross sections for (c, d, and e) indicated. f is beginning to reveal molecular structure. Reproduced with permission from Vater, W., W. Fritzsche, A. Schaper, K. J. Bohm, E. Unger and T. M. Jovin (1995). Scanning Force Microscopy of Microtubules and Polymorphic Tubulin Assemblies in Air and in Liquid. *J. Cell Sci.*, **108**(Pt 3): 1063–1069.

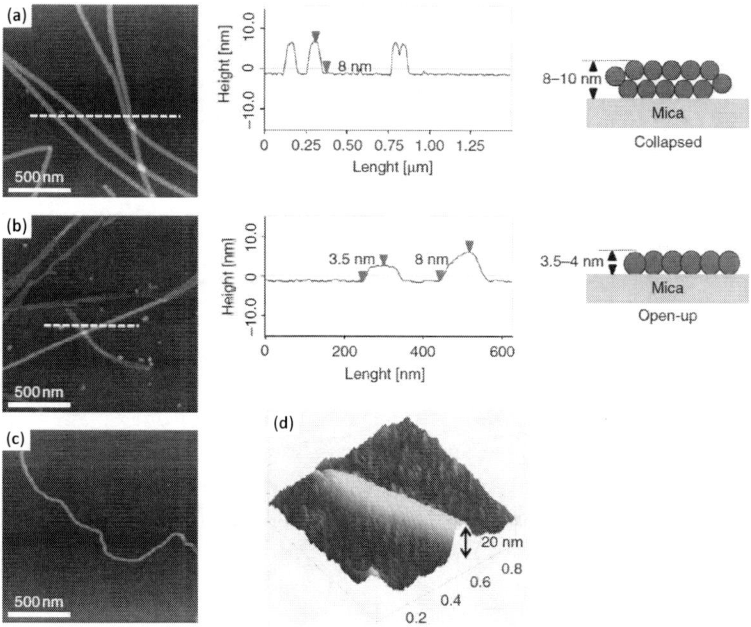

Figure 4.3 (a) Collapsed (b) splayed (c) gluteraldehyde fixed and (d) liquid-stabilized microtubules imaged with atomic force microscopy. Hamon, L., P. A. Curmi and D. Pastre (2010). High-Resolution Imaging of Microtubules and Cytoskeleton Structures by Atomic Force Microscopy. *Methods Cell Biol.*, **95**:157–174.

4.1.4 Cell Membranes

An understanding of membrane mechanics is crucial in determining how biological membranes respond to both external and internal stimuli. Such stimuli include shear stresses, e.g., stress imposed by blood flow onto an endothelial cell that lines a major artery (Chen et al., 2008; Li et al., 2008a), internal pressure changes, e.g., pressure generated inside a cell during cell aspiration or injection (Henriksen and Ipsen, 2004; Ohashi et al., 2006), and bending and stretching, e.g., deformations due to a red blood cell squeezing through a capillary, or due to a cell or liposome being probed by an atomic force microscope or injection device (Cevc et al., 2002a; Chen et al., 2007; Brochu and Vermette, 2008).

Biological membranes, like any other material, can deform in the three fundamental modes of bending, stretching, and/or shearing. However, the primary way in which these membranes actually store energy elastically is in bending (Gompper and Kroll, 1995b; Hotani et al., 1999), whereas in-plane shear deformation generally dissipates energy in a fluid-like manner. When subjected to bending, the outer surface of the phospholipid membrane bilayer is loaded in tension, and the inner surface is loaded in compression, with the molecular interface between the two layers essentially serving the role of the neutral axis or neutral plane.

Despite the ability of continuum theories to describe cell membrane deformation modes, cell membranes are of course inhomogeneous structures that contain sterols, ion channels, and other classes of membrane proteins. The type and quantities of these membrane components varies among different species and cell types, and therefore the mechanical properties of the membranes will change depending on the cell type. It has been suggested by Hulbert (2003) that the physical characteristics of membrane lipids are one of the main defining factors for how cells perform their specific functions and maintain viability.

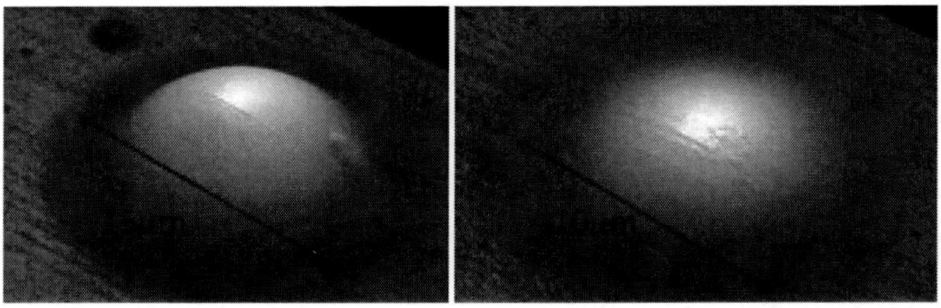

Figure 4.4 Atomic force microscopy image of laboratory-prepared liposomes. These liposomes, which are essentially single cells devoid of their living components, were prepared and deposited on a glass slide in vacuum. They were then hydrated with PBS and refrigerated for 24 h. The liposomes were then imaged within a meniscus of PBS in the contact mode using a Veeco DNP-S chip with the most compliant sampling tip $k = 0.01$ N/m. Efforts were made to reduce deformation of the liposome from the sampling stylus by employing a rapid sampling rate with light pressure.

This unique structure has been described as behaving like a "two-dimensional fluid" with the two layers of the membrane flowing freely past each other and with individual phospholipid molecules moving freely within the membrane and exchanging locations across the membrane (Dai and Sheetz, 1995; Fygenson et al., 1997). Given these membrane characteristics, the two Poisson's ratios of this transversely isotropic material have been reported as 0.49 (Charras et al., 2004). This ratio is consistent with the claim that the cell membrane is incompressible (Hamill and Martinac, 2001). The ability of the phospholipids that compose the membrane to flow freely over each other indicates that the membrane has a low or zero in-plane shear modulus (Rahimi and Arroyo, 2012). It is these physical properties that make physical characterization a challenge. However, isolation and imaging via AFM is relatively straightforward (Fig. 4.4).

The area expansion modulus has been used to describe the mechanical properties of a biological membrane, and is defined as the tension that is required for a unit increase in strain. For phospholipid bilayers, area expansion moduli have been shown to be on the order of 16–240 mN/m (mJ/m^2). The area expansion modulus of the membrane is related to the bending modulus, κ_B, used by Atilgan et al. (2006) through

$$\kappa_B = \frac{K_A t^2}{24}, \tag{4.1}$$

where K_A is the area expansion modulus of the membrane and t is the thickness of the membrane (Evans and Rawicz, 1990; Boal, 2002).

Evans and Rawicz (1990) have shown that the area expansion modulus decreases as the degree of saturation of the lipids decreases. Unsaturated bonds will cause lipid chain-packing irregularities and increase chain and membrane flexibility (Lehninger et al., 2005). Further, shorter-chain fatty acyl groups cause chain packing irregularities and increase membrane flexibility (Lehninger et al., 2005). This has the opposite effect of adding cholesterol or other membrane components to the membrane. Additional membrane components tend to decrease membrane flexibility (Evans and Rawicz, 1990).

When looking at the images of Figs. 4.1, 4.2, and 4.4 in succession, it is remarkable that each of these three structures, collagen, with its long, sticky fibrous nature, tubulin, with its girder-like structure, and the cell membrane, with its graceful yet tough curves, have found their respective paths into the fabric of life.

4.2 Atomic Force Microscopy

We now examine the role that atomic force microscopy has played in enabling visualization of key proteins and key organellar and subcellular structures. The image produced by AFM is constructed line-by-line and the data is sent to a "scan trace" window where the operator essentially monitors the error between the forward and backward scan directions. In the absence of mechanical vibration and "thermal drift," these two lines would lie directly on top of each other. However, because of varying coefficients of friction as the sample dries and because of the unequal angles at which the probe contacts the surface in the two directions (Wang et al., 2003), these traces are not identical and the user typically tunes the instrument by adjusting gains to minimize the error between the two signals. In a sense the enhanced accuracy, which brings the idealized or recorded state of the sample into close approximation with the physical state, costs only time but not additional energy once error minimization has been achieved. More precisely, the energy required to hold the instrument at the optimal gain does not cost

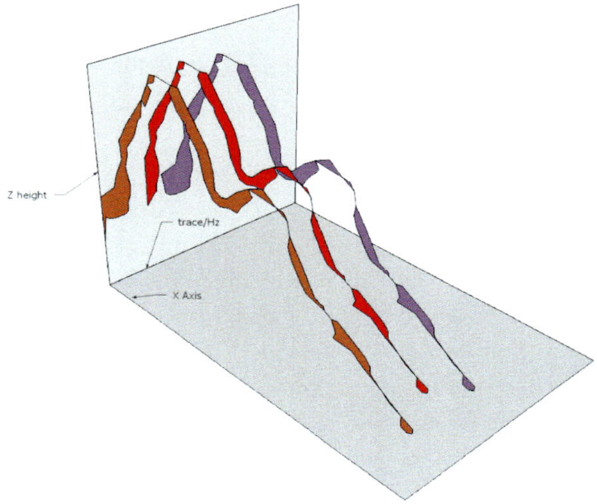

Figure 4.5 This graph shows a scan trace projected over time. Note the data collected from the trace and retrace lines of the scan form the composite shown in the image screen. Slight changes in the line-by-line scan create a surface topography by designating Cartesian coordinates in three axes to a grid of data points.

any more energetically than holding it at the countless other suboptimal gains available to the user. The traces depicted in Fig. 4.5 depict three trace-retrace scans of a typical image. Notice that the scanner reached its threshold at the flat peaks near the left of the figure.

As the stylus rasters over a sample, data is collected at a constant frequency. Figure 4.6 illustrates how steep regions may receive a greater number of bits per unit horizontal distance than flatter regions.

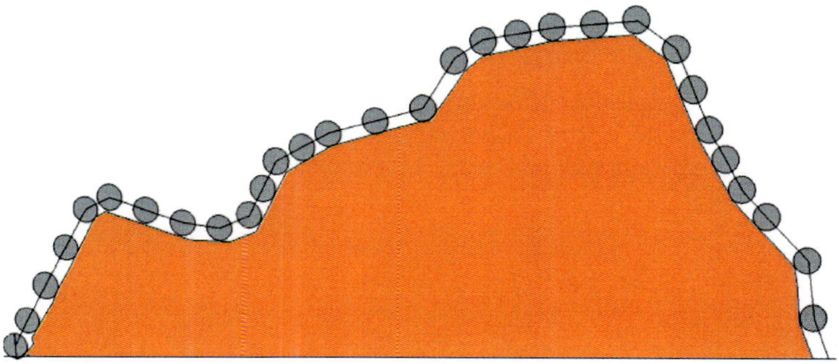

Figure 4.6 This graph represents the series of data points collected by the scan trace as the stylus passes over the sample. The coordinates of these points form the basis of a mathematical model for digitally imaging a surface one line at a time.

The accuracy of this data lies in correctly adjusting and monitoring the inputs and feedbacks of the machine. In an ideal situation, a stylus tracks back and forth across a surface, and records the same information in each direction. However, as mentioned above, apparent sizes and dimensions must be considered with respect to the tip geometry of the stylus (Figs. 4.7, 4.8). Another artifact that AFM is notorious for is so-called "convolution" (Fig. 4.7). While the output may yield an apparent width W in an image collected in an AFM using a stylus with tip radius R, the actual lateral size is L. This relationship is given via

$$L = W - 2\sqrt{2Rh - h^2} \tag{4.2}$$

A similar phenomena occurs with circular cross sections, where the apparent radius is given as

$$r = \frac{W^2}{16R}. \tag{4.3}$$

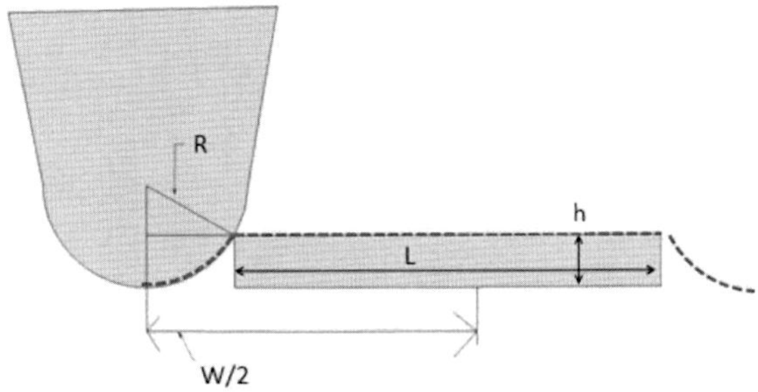

Figure 4.7 Samples with rectilinear geometry and height less than or equal to the radius of the stylus show an apparent profile contour similar to the dashed red line.

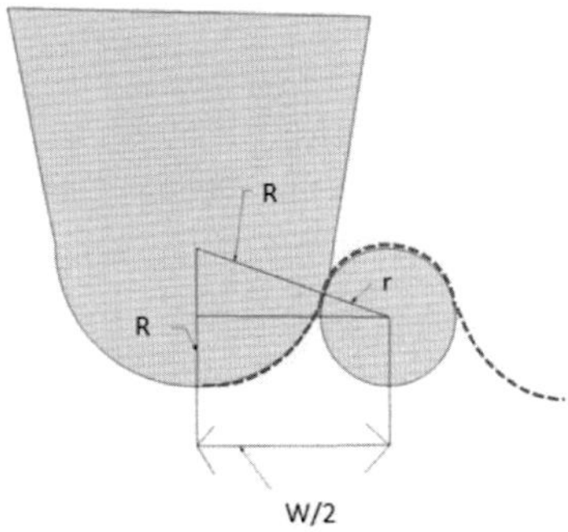

Figure 4.8 Samples with spherical and radiused geometry in which the sample radius is less than or equal to the tip radius display contours of this nature (shown in red).

Imaging issues can be resolved largely by adjusting the gains, cantilever height, scan rate and voltage setpoint. For example, a set of regular vertical or diagonal lines begins to appear in an otherwise clear image. This indicates that the stylus is leaving the surface at intervals and "ringing" as it passes through the air. The forces driving the stylus up are greater than the forces holding the stylus in contact with the sample. The user must increase the force of the cantilever, or reduce the scan speed. When sampling soft tissues or delicate samples, increasing the gain can deform, damage or destroy the sample Fig. 4.9, while on more rigid surfaces, adjusting the gains too high will result in "ringing" in the stylus Fig. 4.10. Reducing the velocity of the tip as it tracks the surface will decrease the forces driving the cantilever back up.

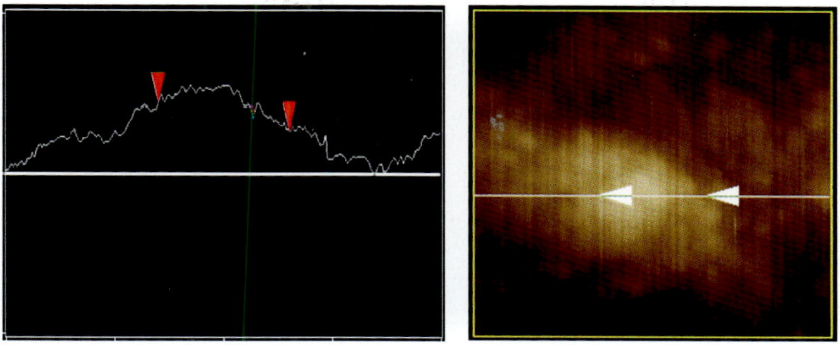

Figure 4.9 A section reveals a specific arbitrary horizontal line of data points in the direction perpendicular to the scan direction (vertical) much the same as a single line represented in the scan trace. This particular sampling is marred by a recognizable artifact, seen here as vertical lines running parallel across the sample. In the image at left, we can see that the lines (shown on the trace as spikes of similar size) are distributed fairly evenly over the otherwise definable topography of the sample.

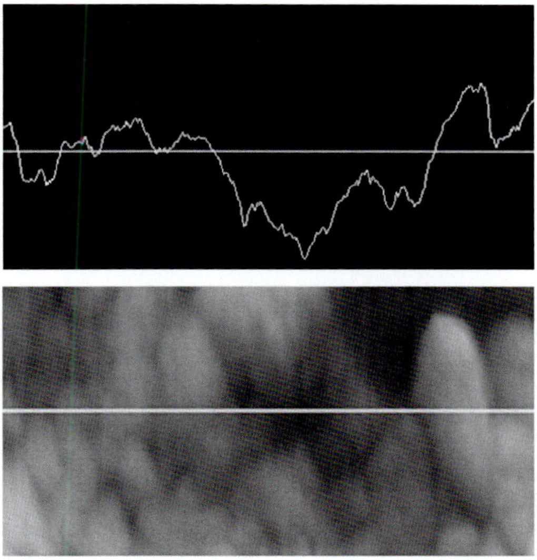

Figure 4.10 Images were produced in the same manner as the images in Fig. 4.6. However, in this case, the sampling was taken at a slower scan rate with a higher gain.

Keeping the stylus in contact with the surface is crucial for getting a clear and accurate image. Even when the optimum conditions are met, there will still be artifacts present Fig. 4.8. Recognizing the differences between features and artifacts is critical to the interpretation of AFM images.

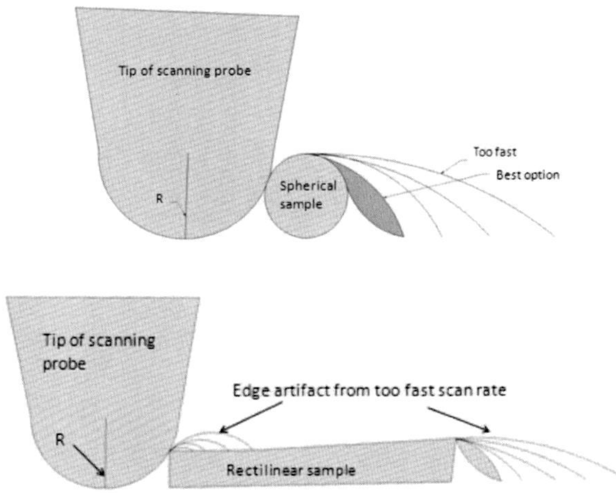

Figure 4.11 Surface topography error caused by scan rate too high; note the shaded area, which shows optimal lateral speed range.

4.2.1 Air Tapping

In this mode, the cantilever is vibrating vertically near (typically just below) its fundamental resonant frequency, contacting the sample in a series of repeated vertical "pecks." This technique is applicable to any dry sample with a vertical surface topography range of less than 5 μm.

If a "false engage" occurs (Fig. 4.12), the stepper motor approach interval may be reduced to 0.5 μm, and the data scale set to its greatest setting. The tip may then be stepped down several times until the trace begins reading a surface. Depending upon the machine and the backlash in the drive screw of its stepper motor, this may take several adjustments before actual engagement occurs. After first contacting the surface, the stepper motor interval should be reduced to its smallest setting, and the approach continued until the Z-height meter reaches the middle of its range. At this point necessary adjustments to the gains and voltage setpoint may proceed.

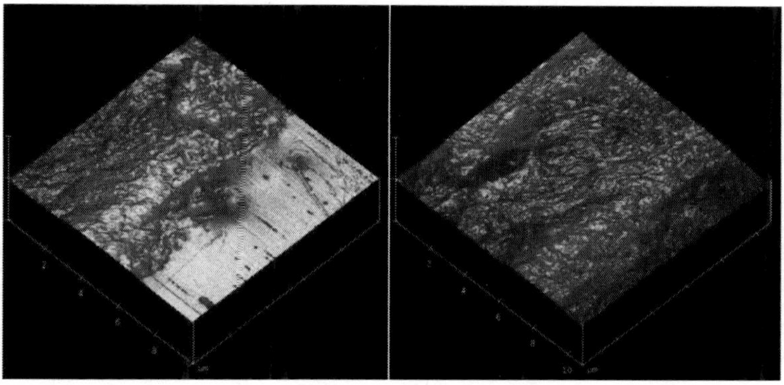

Figure 4.12 In this pair of 10 μm images, the left image is a false engage of an *Eragrostis* leaf surface. After withdrawing, re-tuning, and re-engaging the stylus, the right image of the same area was obtained.

One of our recent investigations has been into the role that conformational changes in the cell wall structures of drought-resistant and drought-tolerant plants may play in resisting hypo-osmosis and desiccation (Layton et al., 2010). Understanding the scale and topography of the sample area is important to recognizing usable data. For example, in *Eragrostis*, on the upper leaf surface, there is a series of ridges, that when sampled, at first appear as a series of mountaintops appearing through a fog layer. A re-engage and adjustment of the feedbacks improved the resolution.

Assigning a data scale that corresponds to the scan area is critical. On rough samples such as the *Eragrostis* leaves shown in Fig. 4.12 there is a much greater chance of encountering objects greater than 5 μm high when the scan area has a linear dimension greater than 5 μm. A scan of 1 μm^2 is less likely to contain vertical features greater than 5 μm.

4.2.2 Contact Mode

Our recent investigation of the cellular structures and mechanics of the drought tolerant fern *Polypodium polypodioides* presented a number of challenges (Layton et al., 2010). We were attempting to collect micro and nanoscale images of the samples in both the hydrated and dry conformations. However, these samples undergo significant physical deformation as a part of this wetting and drying process. A fresh sample would change shape when placed on a glass slide, faster than we could image it. We combated this by applying a meniscus of PBS over the sample. The necessity of using a wet sample ruled out air tapping, and we changed to a Veeco DNP-S cantilever tip and ran with the Veeco contact AFM protocol.

4.2.3 AFM of DOPC-DOPS Liposomes

The phospholipids of the cell membrane, the nucleus, and cell organelles literally "hold life together." All of the genetic information and cell metabolic machinery relies on the maintenance of a barrier between the cellular contents and the external milieus. Phospholipid mass alone comprises 24% of liver, 17% of red blood cell, and 39% of mitochondrion membranes (Erickson et al., 1937). Approximately 5–15% total mass in most animals is membrane and in humans the total cell and organelle membrane has an approximate area of 1 km^2. Active or passive transport occurs across these membranes. Thus as discussed above, the mechanical safety factor is paramount to cell and thus organism survival. In fact, all membranes are in a sense "inherited" from parent cells (Cavalier-Smith, 2001), giving an intriguing perspective to the concept of the continuity of life existing within a protective barrier passed on from generation to generation.

We recently imaged phosphatidylcholine-based liposomes (PC) because of their prevalence and importance to eukaryotic cells and their organelles. The selected liposomes DOPC and DPPC have differential rigidity above and below 37°C, DOPC being more liquid-like and DPPC more gel-like. DPPC in particular is the main constituent of the pulmonary surfactant, a mixture of lipids and proteins that control the surface tension of the fluid lining of the lung, allowing its rapid expansion and compression during breathing.

In nature, cells have different strategies for cell membrane maintenance. For example in red blood cells, a layer of spectrin maintains cellular integrity (Lang et al., 2007). In neurons, the extremely long and articulated membrane provides and organism with the ability to rapidly coordinate sensory and motor control throughout all organs and limbs (Kandel et al., 1991). This same balance is critical for understanding the mechanics of individual liposome diffusibility and stability (Gompper and Kroll, 1995a; Cevc et al., 2002b).

Recent efforts to characterize liposomes with AFM include Ruozi et al. (2005) and Ruozi et al. (2007), who observed the aggregation of egg-yolk phosphatidylcholine (PC) dimethyldioctadecylammonium bromide (DDAB) liposomes over a seven-month study using non-contact AFM. Most recently, Brochu and Vermette (2008) used a finite-sample thickness corrected Hertz-contact model to measure a Young's modulus approximately 3 kPa for surface-bound liposomes. Zhao et al. (2007b) used a simple, elegant experiment to measure the Young's modulus and critical buckling length of lipid tubules. In a study to test the efficacy of liposomes as a cisplatin delivery system, Ramachandran et al. (2006) found the drug-loaded 250 nm structures to be nearly twice as stiff as water-filled controls perhaps due to the embedment of the ringed hydrophobic cisplatin within the bilayer. Also of recent interest is a high-profile project from MIT designed to optimize RNAi delivery via liposomes (Akinc et al., 2008; Anderson, 2008).

There are a growing number of studies in the field of atomic force mapping of liposomes (Rhee et al., 1998; Quist et al., 2000; Almqvist et al., 2004; Ramachandran et al., 2006). Previous work in examining the relationship between molecular structure and mechanical behavior include (Bouffioux et al., 2007). However, this group only studied the system in a monolayer form. Other efforts to measure the mechanical properties of phospholipid structures include those of Rhee et al. (1998) who found amyloid β protein vesicles to be stiffer than controls in the presence of Ca^{2++}. In the work of Rhee et al. (1998), however, a Hertz model for sphere–sphere contact was used and no actual physical values were reported, only stiffnesses normalized to controls.

There are several existing methods for predicting membrane shapes under neutral load. These have been categorized by Bloor and Wilson (2000) as (a) solving the Euler-Lagrange equations for the integral expression for the total surface energy (Majhenc et al., 2004); (b) minimizing a parameterized model for the shape of a vesicle (Svetina and Zeks, 2002; Hotani et al., 2003); and (c) finite element analysis (Feng and Klug, 2006; Allen et al., 2009b). Multiscale models for predicting micelle or liposome formation or membrane-protein interactions, e.g. (Scott et al., 2008) using molecular assembly mechanisms have emerged as a promising way to bridge the atomistic to cellular scale (Cooke and Deserno, 2006; Arai et al., 2007; Michel and Cleaver, 2007).

4.3 Further Considerations

In this chapter we have introduced a new set of concepts and a few practical equations for how we might move toward a science wherein biology, physics, and computational mechanics are may be discussed under a single framework of knowledge. We now move into a more general discussion of some recent papers wherein nanoscale imaging is used to digitize the engines of life. It is worth stating here that when Crick and Watson resolved the structure of DNA, what they did was to use an electronic system X-ray crystallography to take a picture of physical reality (Watson and Crick, 1953). With their seminal effort, they were able to seine this subtle and elegant structure for reproduction from the inner sanctum of the cell and bring it into the noosphere where it can be copied and understood innumerable times. If you are to read through Dawkin's recent compendium of the great works of the 19th century (Dawkins, 2008), you will read several essays in which people become highly regarded for performing similar tasks, namely, taking the apparent messiness of physical reality and distilling it into a simple set of equations or statements. The next few decades are likely to see humanity capitalizing on similar acts whereby we push the envelope of what Dyson calls "the end of Darwinian evolution" (Dyson, 2007). Specifically, we will see people build artificial living systems that are capable of meeting the criteria for being alive listed early in this chapter, but which will likely be simpler in the same way that a jet's engine is simpler than a bird's wing. Whether or not these new species will out-compete, naturally evolved systems will be answered only through observation, but it is likely that any artificial living systems that are closer to the "left wall" will be more successful than those closer to the right asymptote.

4.3.1 Imaging the Engines of Life

Recently, Bussiek et al. (2007) used AFM to image individual chromatin strands from African Green Monkey DNA. The AFM images are of individual chromatin strands fixed to a poly-lysate layer on a mica slide. The concern of this paper is less with the overall structure of the nucleus and more on the distribution of nucleosomes along alpha satellite DNA strands. While the AFM images are of high quality, they are not of complete nuclei, but the reconstituted ingredients thereof. This is a good example of where we are with the ultimate confluence of atomic force microscopy and other techniques such as NMR, microcomputer tomography, and X-ray crystallography. While AFM is capable of quickly resolving the surface structure of an isolated biological structure, it is not yet capable of penetrating the inner topology of biomolecules in the same way that NMR, μCT and X-ray are. Eventually technologies will emerge where the relatively simple sample prep of AFM will meet the 3D resolving power of NMR, μCT and X-ray technologies. Our own group has made small steps in this direction (Layton and Buzby, 2010).

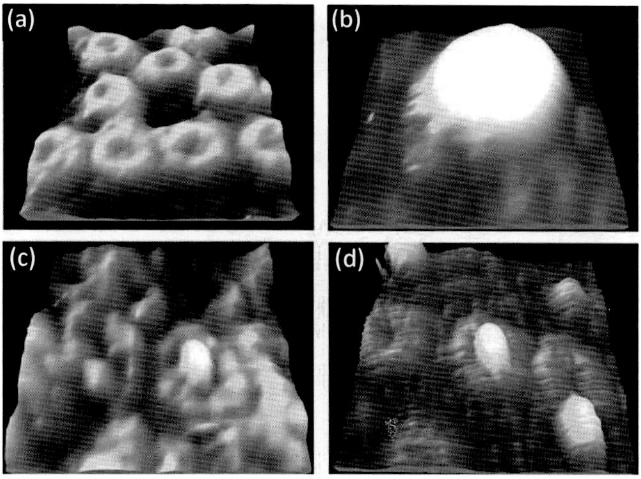

Figure 4.13 AFM images (each 550 nm) of the cytoplasmic (a–c) and nucleoplasmic (d) faces of the nuclear envelope of Xenopus laevis oocytes displaying sequential steps of herpes simplex virus type 1 (HSV-1) genome delivery into the nucleus through nuclear pore complexes. a: Nuclear pores before exposure to HSV-1 capsids enclosing the viral genome. b, c: The intact, genome enclosing HSV-1 capsid binds (b) to the cytoplasmic face of the nuclear pore and injects (c) the bulky viral genome as a condensed rod-like structure into a remarkably enlarged nuclear pore central channel. d: Genome translocation through the nuclear pore central channel into the nucleus. Modified from Shahin, V., Hafezi, W., Oberleithner, H., Ludwig, Y., Windoffer, B., et al. (2006). *J. Cell Sci.* 119:23–30 Reprinted with permission of *Journal of Cell Science.*

In another paper of interest in how the nuclear envelope degrades during programmed cell death or apoptosis, Kramer et al. (2008) explored apoptosis of the nuclear envelope, specifically the nucleocytoplasmic nuclear pore complexes. The paper includes high-quality AFM imagery of the nuclear envelope, both in height mode and as a function of force/stiffness. The nuclear pores as resolved by e.g. Fig 4.13 & 4.14 are very clearly defined and they are able to illuminate physiological changes to the lamina and pore complexes in the process of cell degradation. Within the context of the first section of this chapter, we see how if the cell is willing to invest the energy to dismantle itself gracefully, the neighboring cells are not encumbered by the toxin release that would occur in the wake of a catastrophic necrosis of the cell. Thus, apoptosis can be regarded as another example of how an organism shields itself from entropy at the cellular scale.

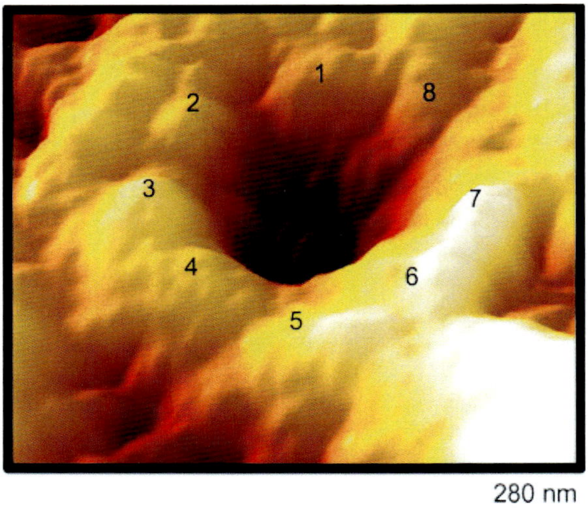

Figure 4.14 Atomic force microscopy image of an individual (representative) giant pore detected on the cytoplasmic nuclear envelope surface of Xenopus laevis oocytes 11 min after dexamethasone (glucocorticoid) injection into oocytes. A single giant pore is 300 nm in diameter and exhibits an eightfold symmetry (indicated by numbers) known to be characteristic for 'regular' nuclear pore complexes. Modified from Kastrup, L., Oberleithner, H., Ludwig, Y., Schafer, C., Shahin, V. (2006). *J. Cell Physiol.* **206**:428–434 reprinted with the permission from Journal of Cellular Physiology.

The nuclear pore complex is perhaps among the most "unlikely" structures in nature. Perhaps not surprisingly, the majority of the opponents of Darwinian Theory lack access to the microscopes and academic training required to produce the images produced by science with some of the more powerful imaging techniques developed within the last century. The nuclear pore complex, which is built from a large number of individual proteins, is responsible for regulating material transport to and from the nucleus, or the inner sanctuary of life's blueprints. Lim and Fahrenkrog (2006) present several images of the nuclear pore complex taken with scanning force microscopy. This paper is largely concerned with imaging the constituent parts of the functioning nuclear pore complex.

Another step that has been taken toward integrating nanoscale imaging with predictive mechanics exists in the paper of Pesen and Hoh (2005). In their paper, they were discerning the micromechanical architecture of the endothelial cell cortex. The authors employ AFM imagery to identify and define micromechanical elements of the cell. They also use CFM images for comparison. While there are no nucleus-specific images, there are several good AFM techniques presented in this article relating to interpreting the AFM images with models.

Atomic force microscopy is also frequently used in the mode that the name implies. For example, Dahl et al. (2005a,b) have begun a series of studies concerned with interpreting the mechanical properties of isolated nuclei and their viscoelastic mechanical properties. While the only images of the nuclei are optical, the authors do use the AFM to interrogate for the purpose of extracting mechanical deformation information. In a similar paper, Moreno-Herrero et al. (2005) employ AFM to interpret chromosome function as a function of mechanical properties. Finally (Shahin, 2006) (Figs. 4.11 and 4.12) contains some of the highest quality AFM images of the cytoplasmic nuclear membrane and nuclear membrane pore architecture taken to date. Perhaps in the near future, the high-resolution of AFM will be combined with the three-dimensional imaging capabilities of technologies such as MRI, μCT, and confocal microscopy so that a more robust and real-time picture of nanoscale cellular biomechanics will come into focus.

Problems

4.1 Explain the difference between stiffness/compliance and strength/weakness.

4.2 Explain how the stepper motor differs from the piezomotors of an atomic force microscope.

4.3 What are the primary roles of (a) collagen, (b) tubulin, and (c) phospholipid bilayer in an organism?

4.4 Using the equipartition theorem of Chapter 2, find the thermal noise present in an AFM cantilever beam with a spring constant of 0.1 N/m at 300 K. How many atomic diameters of noise is this approximately? (Assume the atomic diameter to be 1 Å).

Chapter 5

Cell–Substrate Interactions

Rebecca Lownes Urbano and Alisa Morss Clyne

Mechanical Engineering and Mechanics, Drexel University, 3141 Chestnut Street, Philadelphia, PA 19104, USA

asm67@drexel.edu

5.1 Introduction

Like many engineered structures, most cell types attach to a foundation that provides adhesion sites and mechanical support. This foundation, called the extracellular matrix, communicates a variety of biochemical and biophysical signals for essential cell functions including survival, proliferation, migration, and assembly into a functional tissue. The extracellular matrix is a protein mesh composed of structural proteins such as collagen, elastin, laminin, and fibronectin, as well as signaling and storage molecules such as proteoglycans (Fig. 5.1). Extracellular matrix composition and structure vary across different tissues and in health and disease.

Molecular and Cellular Biomechanics
Edited by Bradley E. Layton

ISBN 978-981-4316-83-5 (Hardcover), 978-981-4613-44-6 (eBook)
www.panstanford.com

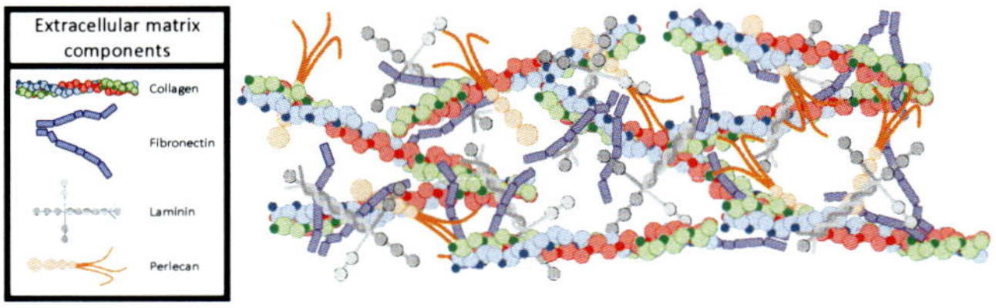

Figure 5.1 The extracellular matrix is a protein mesh surrounding cells that provides both mechanical and biochemical support. While matrix content varies with tissue and disease, certain core molecules are consistently present. Structural proteins such as collagen, fibronectin, and laminin define matrix mechanical properties and cell adhesion, whereas proteoglycans such as perlecan store signaling molecules like growth factors.

The extracellular matrix, together with the cells embedded in it, defines tissue mechanical properties, specifically the elastic modulus E or stiffness, k. Normal tissue moduli range from 1 kPa for soft tissue such as liver, to 10–100 kPa for intermediate tissues such as arteries, and more than 100 MPa for hard tissues such as bone. Early in vitro studies suggested that substrate modulus may affect cell function; however, it was difficult to independently modulate substrate biochemistry and mechanics. In 1997, Pelham and Wang used polyacrylamide gels coated with collagen to determine substrate stiffness effects on cell spreading and focal adhesion formation (Pelham and Wang, 1997). Polyacrylamide gel stiffness can be changed by varying the acrylamide-monomer to bisacrylamide-crosslinker concentration; gels do not adsorb proteins so specific matrix proteins can be linked to the gel surface to control the cell adhesion ligand; gels have superb optical quality that permits high resolution immunofluorescent microscopy; and the materials are familiar to biology and bioengineering laboratories that perform electrophoresis. With this new model system, the study of substrate stiffness effects on cell function rapidly expanded. Cell types ranging from endothelial cells to neurons to mesenchymal stem cells have been studied on gels of stiffness ranging from 15 Pa to 700 kPa.

Cells sense substrate properties via integrins, transmembrane heterodimeric proteins that attach extracellularly to specific matrix protein domains and intracellularly to the actin cytoskeleton (Fig. 5.2) (Barczyk et al., 2010). The 24 integrins are formed of different combinations of 18 α and 8 β subunits, with both subunits contributing to integrin–ligand specificity. For example, $\alpha_1\beta_1$ and $\alpha_2\beta_1$ integrins bind to the GFOGER sequence in collagen. $\alpha_v\beta_3$ and $\alpha_5\beta_1$ bind to the RGD sequences of vitronectin and fibronectin, respectively. Integrin–matrix interactions that depend on integrin and matrix protein density as well as the specific integrin–matrix protein pair control many cell processes. Integrins assemble into focal complexes that can mature into larger focal adhesions, which contain a plethora of linking and signaling proteins. Proteins such as vinculin and talin anchor actin stress fibers to provide a mechanical connection, whereas other proteins such as focal adhesion kinase provide biochemical signaling. When external forces are applied, cells exert traction forces on the substrate to which they are attached via focal adhesions. Cell traction forces deform softer materials, and the substrate resists these cell-generated forces if it is sufficiently stiff to prevent deformation (Janmey and McCulloch, 2007).

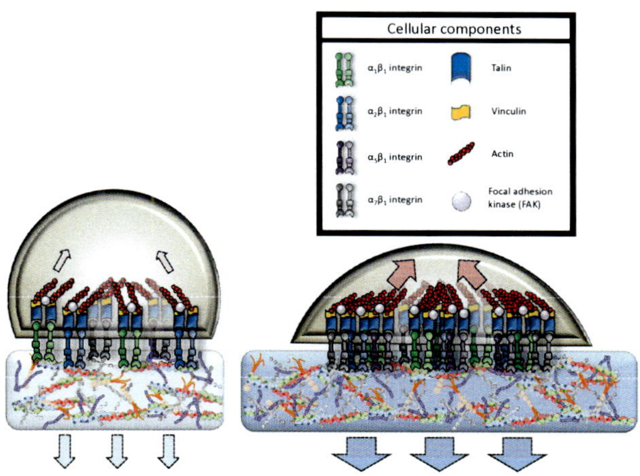

Figure 5.2 Cells sense substrate stiffness through integrin-extracellular matrix linkages. Cell-surface integrins bind specific extracellular matrix components and transmit mechanical and biological signals into the cell. Anchoring molecules, such as talin and vinculin, provide structural support within focal adhesions while other molecules, such as focal adhesion kinase, are integral to intracellular signaling. Soft substrates (left) provide less resistance to cell traction forces, resulting in decreased intracellular tension and cell spreading. As substrate stiffness increases (right), cell spreading and migration increases due to generation of higher intracellular tension and traction forces.

Many studies investigated the separate effects of matrix ligand and substrate stiffness on cell function. More recently, cell response to changes in both matrix properties has begun to be examined and shown fascinating interaction. These experiments are increasingly physiologically relevant, since matrix ligand density and type as well as matrix stiffness change in diseases such as hypertension, cancer, and diabetes. In this chapter, we review how matrix ligand and substrate stiffness individually and together modify cell morphology, motility, and mechanics. Finally, we introduce applications of these data in understanding pathophysiology.

5.2 Effect of Substrate Stiffness and Matrix Ligand on Cell Morphology

For most cell types, cell-spread area increases with substrate stiffness. In a seminal paper by Pelham and Wang (1997), 3T3 fibroblasts and normal rate kidney epithelial (NRK) cells grown on collagen-coated polyacrylamide gels showed increased cell spreading at higher substrate stiffness. Following this study, other cell types similarly displayed increased cell spread area with substrate stiffness, including endothelial cells (Yeung et al., 2005b; Califano and Reinhart-King, 2008), vascular smooth muscle cells (Engler et al., 2004a; Polte et al., 2004; Zaari et al., 2004; Peyton and Putnam, 2005; Peyton et al., 2006), chondrocytes (Genes et al., 2004), cardiac myocytes (Engler et al., 2008), and astrocytes (Moshayedi et al., 2010). While previous data suggested that non-attachment-dependent cells such as neutrophils might be unresponsive to changes in substrate stiffness (Discher et al., 2005; Yeung et al., 2005b), more recent data suggest that they may also increase their cell area on more rigid substrates (Jannat et al., 2010).

However, cell morphological response to substrate stiffness is not always straightforward. For instance, four human vascular endothelial cell subtypes—umbilical vein (HUVEC), aorta (HAEC), saphenous vein (HSaVEC), and dermal microvascular (HmVEC)—were grown on polyacrylamide gels with elastic moduli of 25, 50, and 75 kPa. To varying degrees, the three venous cell subtypes (HUVEC, HSaVEC and HmVEC) showed improved cell attachment and spreading on the softer 25 kPa gel, whereas the arterial cell subtype demonstrated better cell attachment and spreading on the 50 kPa gel (Wood et al., 2011). Substrate stiffness also affects cell-type specific morphologies. Primary neurons increased neurite branching with decreasing substrate stiffness, while glial cell and dorsal-root ganglia neuron co-cultures showed maximum branching at an intermediate stiffness (Flanagan et al., 2002; Cheng et al., 2011).

In addition to substrate stiffness, the matrix ligand amount affects cell morphology. In 1990, Ingber tested the effect of matrix ligand density on cell spreading and proliferation using bacteriologic dishes precoated with different fibronectin densities as well as soluble RGD peptides to inhibit integrin binding to fibronectin. Endothelial cells increased in spread area and DNA synthesis as the fibronectin surface density increased from 250 to 10,000 molecules/μm^2 (Ingber, 1990). In later experiments, micropatterned substrates were used to control cell spreading and the distance between focal adhesions. These data confirmed that cell shape, which is affected by matrix ligand density, is a critical factor controlling cell switching from growth to apoptosis (Chen et al., 1997). Similarly, when fibroblasts were seeded onto substrates with different type I collagen surface densities, cell area initially increased with ligand density. However, at a ligand density close to the cell surface integrin density, cell area declined. This biphasic response could be related to decreased traction force per integrin–matrix protein bond as ligand density increases (Gaudet et al., 2003). The initial cell spreading rate (up to 4–6 h after cell seeding) also increased with matrix ligand density. When primary rat hepatocytes were seeded onto substrates coated with laminin, fibronectin, or types I or IV collagen, cell spreading rate increased from 50 to 740 μm^2/h as coating density increased from 1 to 1000 ng/cm^2. Both cell spreading rate and cell spread area were independent of matrix ligand type (Mooney et al., 1995).

5.3 Morphology: Integration of Biochemical and Biophysical Factors

Substrate mechanics and biochemistry (specifically ligand density and type) interact to vary cell morphology in new ways. Smooth muscle cells (SMC) grown on soft 1 kPa gels showed maximum cell spread area at an intermediate collagen density (approximately

100 ng/cm^2). When gel modulus was increased to 8 kPa, the collagen concentration for maximum cell spreading increased to between 100 and 1000 ng/cm^2 (Engler et al., 2004a). Similar results were observed for astrocytes grown on substrates of different stiffness with increasingly bound poly-d-lysine (Moshayedi et al., 2010). Interestingly, primary human aortic smooth muscle cells cultured on fibronectin-coated substrates showed increased spreading at 8 versus 0.8 μg/cm^2 surface density on soft polyacrylamide gels but not on hard polystyrene (Peyton and Putnam, 2005). These effects may be substrate specific, since for hepatocytes, substrate stiffness differently modulated ligand density effects on cell spreading. On stiff substrates, fibronectin concentration had a large effect on cell area, while varying fibronectin concentration on soft substrates had little effect (Semler et al., 2005).

In addition to cell area, other morphological features are also affected by the interplay between biomechanical and biochemical factors. When bovine aortic endothelial cells (BAEC) were grown on gels coated with 100 μg/mL type I collagen covering a range of elastic moduli, inter-cell network assemblies were only observed on lower stiffness gels (0.2 to 1 kPa). When collagen concentration was decreased to 1 μg/mL, network assembly was then present on gels with higher moduli (2.5 to 10 kPa) (Califano and Reinhart-King, 2008).

5.4 Effect of Substrate Stiffness and Matrix Ligand on Cell Motility

Similar to cell morphology, cell motility is also affected by substrate stiffness. Lamellipodia activity and motility of rat kidney epithelial and 3T3 fibroblastic cells decreased with increasing collagen-coated polyacrylamide gel stiffness, perhaps due to higher focal adhesion stability (Pelham and Wang, 1997). In a subsequent study, a stiffness discontinuity was introduced in the polyacrylamide gel center by varying the crosslinker concentration. Fibroblasts migrated to the stiffer side of the gel when they started on the soft side. However, cells that started on the stiff side of the gel retracted or re-oriented to migrate perpendicular to the boundary when they approached the soft side of the gel (Lo et al., 2000). The authors coined the term "durotaxis" to describe directed cell migration in response to substrate stiffness gradients.

Since these original studies, much investigation has been conducted on substrate stiffness effects on motility of different cell types. Pulmonary arterial endothelial cells (Gray et al., 2003), vascular smooth muscle cells (Wong et al., 2003) and macrophages (Nemir et al., 2010) all migrate toward higher stiffness regions. Neutrophils migrated more slowly but more consistently on stiffer gels, which resulted in a larger net movement (Oakes et al., 2009). However, smooth muscle cell migration speed showed a biphasic response with substrate stiffness, with cells having a maximal migration speed at an intermediate polyacrylamide gel stiffness (Peyton and Putnam, 2005). Recently, the conditions necessary for durotaxis to occur have started to be quantitatively analyzed so that appropriate materials can be engineered for various applications. Kawano and Kidoaki (2011) used elasticity-tunable hydrogels to produce controlled stiffness gradients with predetermined width and magnitude and measured the effect on fibroblast durotaxis. Durotaxis only occurred for an increase in substrate rigidity of greater than 30–40 kPa, a narrow gradient width (<50 μm) roughly equivalent to a single cell attached area, and high stiffness gradient (>30–40 kPa/50 μm) under the conditions tested.

Cells also exhibit haptotaxis, migrating from areas of lower to higher adhesive molecule substrate surface density (Carter, 1967). Cell motility is determined by adhesion strength. When cells weakly adhere to the substrate, they are unable to generate adequate traction forces and therefore cell movements are small. When cells strongly adhere to the substrate, they are unable to dynamically dissociate and reorganize adhesion sites, and therefore cells become immobile. Appropriate intermediate adhesion strength is important for efficient cell migration (Dimilla et al., 1993). As cells migrate up the substrate ligand gradient, they eventually become stably fixed due to high cell adhesive forces. For directional migration, it has been suggested that cells require an asymmetry in cell adhesiveness from the front to the rear of the cell (Dimilla et al., 1991).

Cell adhesion strength depends on both ligand density and ligand-receptor binding affinity, among other factors. Cell adhesion strength increases with ligand coating density (as well as with integrin number and diffusion rate), since more integrin–matrix protein bonds can form (Palecek et al., 1997; Gallant et al., 2005). Cell migration requires both lamellopodia protrusion and trailing edge retraction. At low ligand densities, cell protrusion is the limiting factor. At high ligand densities, cell retraction is the limiting factor (Huttenlocher et al., 1996; Huttenlocher et al., 1997). Matrix protein density can further influence other factors required for effective cell motility via outside-in signaling, including focal adhesion size and turnover as well as actin flow rate (Gupton and Waterman-Storer, 2006).

Intermediate integrin–matrix protein affinity similarly enhances cell migration, as demonstrated using antibodies of varying affinity to the fibronectin integrin (Duband et al., 1991). Huttenlocher et al. (1996) also used integrin activating antibodies to show that integrin activation decreased cell motility on fibrinogen. Part of the difference in cell adhesion strength between the cell leading and trailing edge may be related to increasing integrin affinity at the leading edge (Galbraith et al., 2007). Integrin–matrix protein affinity likely varies with different receptor–ligand combinations. Single-cell force spectroscopy has begun to elucidate strength differences in single integrin–ligand bonds (Lehenkari and Horton, 1999; Li et al., 2003; Taubenberger et al., 2007), and bond affinity regulates adhesion strength (Kuo and Lauffenburger, 1993). Osteosarcoma cells had higher adhesive strength to vitronectin ($\alpha_v\beta_3$ integrin) than type I collagen (β_1 integrins) when ligand density was non-limiting, and endothelial cells showed higher adhesion strength to cyclic as compared to linear RGD sequences (Xiao and Truskey, 1996; Lacouture et al., 2002; Kato and Mrksich, 2004).

5.5 Motility: Integration of Biochemical and Biophysical Factors

While both substrate stiffness and matrix protein density and type impact cell motility individually, these two factors also interact to affect cell migration. Varying the matrix protein concentration can affect cell response to substrate stiffness. Smooth muscle cells migrating on fibronectin-coated polyacrylamide gels showed the highest migration speed at an intermediate substrate stiffness. However, the stiffness required for maximum migration speed shifted with the substrate matrix protein surface density. On substrates coated with 0.8 μg/cm^2 fibronectin, the maximum cell migration speed was achieved on a 51.9 kPa gel. On substrates coated with 8 μg/cm^2 fibronectin, the maximum cell migration speed was achieved on a gel that was two times softer (21.6 kPa) (Peyton and Putnam, 2005). There also appears to be some evidence of interplay between substrate stiffness and chemotaxis. For instance, neutrophils on two-dimensional hydrogels with varying elasticity were more sensitive to the bacterial chemoattractant fMLP on stiffer substrates (Jannat et al., 2010).

5.6 Effect of Substrate Stiffness and Matrix Ligand on Cell Mechanics

Many cell types actively respond to mechanical forces in their environment and produce forces that are essential to their function. Integrins form the mechanical link between the actin cytoskeleton and extracellular matrix proteins. Cells produce traction forces with the substrate, which are required for cell spreading and movement. The contractile force is generated intracellularly by myosin II and F-actin. While early experiments suggested that cells exerted forces on their substrates during movement, these forces could not be measured on typical tissue culture non-deformable substrates such as polystyrene and glass, which have stiffnesses around 3 and 72 MPa, respectively. In a seminal paper, Harris et al. (1980) cultured chick heart fibroblasts on a thin silicone film and observed wrinkle formation as the cells moved. These studies demonstrated for the first time that cells generate traction forces behind the cell leading edge and that these forces can be quantified (~1 dyne/μm in shear) as long as the substrate stiffness is known.

Traction forces are most commonly measured by traction force microscopy (TFM). In this method, the cell-induced substrate deformation is analyzed through the movement of fluorescent beads embedded in the gel. The bead displacement, x, is measured optically before and after the cell applies the traction force. The corresponding force is calculated using the experimentally determined substrate elastic stiffness which can be expressed in terms of the substrate's elastic modulus (Van Vliet et al., 2003). Dembo and Wang (1999) developed an automated computer algorithm to first capture bead marker displacement and then track bead displacement using a correlation-based optical flow algorithm. Once the displacement field is obtained, the traction vector field is constructed including the traction force location, magnitude, and direction (Wang and Lin, 2007).

Cell traction forces vary on substrates of different stiffness. Choquet et al., (1997) pulled on fibronectin coated beads with an optical trap to mimic extracellular matrix resistance. In response to higher pulling forces, cells strengthened the link between the cytoskeleton and integrins, which enabled the integrins to resist stronger forces. Human dermal fibroblasts on hyaluranon and fibronectin substrates of different stiffness showed a higher elastic modulus and more organized actin cytoskeleton on stiffer substrates, which resulted in increased traction forces. Additionally, these cells migrated faster on soft substrates and proliferated more on stiff substrates (Ghosh et al., 2007). Similarly, bovine aortic endothelial cells seeded on polyacrylamide gels of different stiffness coated with collagen demonstrated traction forces that increased significantly from 200 to 400 Pa as stiffness increased from 1 to 10 kPa (Califano and Reinhart-King, 2010). The increase in traction force with substrate stiffness also applies to cells that generate forces, as cardiac myocytes show both increased resting and contractile traction forces on stiffer polyacrylamide gels (Chopra et al., 2011). Interestingly, the maximum contractile force occurred on 10 kPa gels, a stiffness that is similar to native myocardium (Berry et al., 2006; Jacot et al., 2008).

In addition to traction forces, substrate stiffness impacts the ability of cardiac myocytes to generate force. When embryonic quail cardiomyocytes were grown on gels with elastic moduli of 1, 11 and 34 kPa, the highest cell fraction with the organized striations required for force generation was observed at an intermediate modulus of 11 kPa (Engler et al., 2008). In another study, striations did not form in embryonic chick cardiomyocytes on gels at 1 kPa, but did form on 18 and 50 kPa gels and on tissue culture dishes, although striation was not quantified in this study (Bajaj et al., 2010). When contractile work done by beating cardiomyocytes was evaluated on matrices with increasing stiffness, the results formed a bell-shaped curve with maximum work at around 11–17 kPa (Engler et al., 2008). Substrate stiffness has also been shown to affect cardiomyocytes focal adhesion formation, beats per minute, and beating force (Bajaj et al., 2010).

Substrate stiffness affects cell morphology, as discussed previously, as well as cytoskeletal organization. F-actin fiber formation and organization increases with substrate stiffness in fibroblasts (Yeung et al., 2005b; Ghosh et al., 2007) and mesenchymal stem cells (Engler et al., 2006). These changes in turn affect the cell stiffness itself. Fibroblasts matched the gel stiffness up to 20 kPa. This change correlated with more crosslinked F-actin with increasing substrate stiffness (Solon et al., 2007). Cardiac myocytes also increased in stiffness from 4 to 8 kPa as substrate stiffness increased from 0.3 to 30 kPa (Chopra et al., 2011). A similar trend was observed in human mesenchymal stem cells, and the increase in cell cortical stiffness with substrate stiffness was abrogated by the myosin II inhibitor, blebbistatin (Engler et al., 2006).

In 1994, Wang and Ingber showed that increasing cell–matrix contacts by increasing the fibronectin surface density increased both endothelial cell spread area and cytoskeletal stiffness. Their data suggested that the matrix ligand increases cell stiffness by promoting integrin binding to both the matrix protein and the cytoskeleton, which allows forces to be transmitted across these structures (Wang and Ingber, 1994). Similarly, smooth muscle cells seeded on substrates coated with increasing fibronectin density showed both increased resting and contractile force (Polte et al., 2004). Matrix ligand type is important in mechanotransduction, the conversion of mechanical forces to biochemical signals. Endothelial cells on collagen substrates activate $\alpha_2\beta_1$ integrins, while cells on fibronectin activate $\alpha_5\beta_1$ and $\alpha_v\beta_3$ integrins in response to shear stress (Orr et al., 2006), and blocking α_2 or β_1 integrins in cells on collagen inhibited cell morphological changes in response to cyclic strain (Hirayama and Sumpio, 2007).

5.7 Cell Mechanics: Integration of Biochemical and Biophysical Factors

Califano et al. (2010) showed that matrix ligand density increased cell traction force on soft 5 kPa gels only by increasing cell spread area. Their findings suggest that traction force is a function of both cell area and substrate stiffness, and substrate stiffness can alter cell force generation independent of cell area. In cardiac myocytes seeded on either fibronectin and collagen or cadherin coated polyacrylamide gels (to simulate cell–matrix and cell–cell contacts, respectively), myocyte spread area and stiffness increased with substrate stiffness. However, cell spread area and the stiffness increases were consistently higher in cells on N-cadherin as compared to matrix proteins. In contrast, myofibrils were more organized across all gel stiffnesses when cells were on a matrix protein (Chopra et al., 2011).

5.8 Changes in Substrate Stiffness in Disease

The ability of cells to sense the mechanical properties of their environment is pivotal to their proper function within the body (Discher et al., 2005; Reinhart-King, 2011). Changes in extracellular matrix mechanical properties or cellular mechanosensing of the surrounding tissue have been implicated in disease (Fig. 5.3) (Ingber, 2003a; Jaalouk and Lammerding, 2009). In fact, elastography, which uses tissue stiffness measurements to detect disease, is used to diagnose liver fibrosis and breast cancer (Ophir et al., 1991; Bercoff et al., 2003; Sandrin et al., 2003; Nahon et al., 2009).

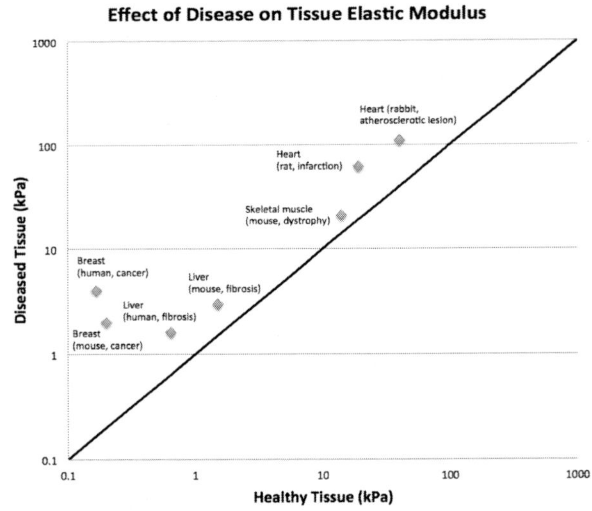

Figure 5.3 Tissue stiffness increases across a wide variety of diseases and tissue types, including breast (Paszek et al., 2005; Levental et al., 2009), liver (Yeh et al., 2002; Yin et al., 2007), skeletal muscle (Engler et al., 2004b), and heart (Matsumoto et al., 2002; Berry et al., 2006).

Cardiovascular disease exemplifies the complexity of studying tissue rigidity. The precise effect of atherosclerosis on arterial mechanical properties is still controversial (Engler et al., 2004b;

Barrett et al., 2009), since results vary based on tissue sample size and location and the testing method. In general, the atherosclerotic lesion elastic modulus is higher than normal tissue (Matsumoto et al., 2002; Barrett et al., 2009). However, mechanical properties vary within plaques, with lower stiffness in lipid-rich areas and higher stiffness in fibrotic areas (Tracqui et al., 2011). Increased stresses within the various plaque regions may be associated with plaque rupture risk, which can lead to myocardial infarction (Ohayon et al., 2008).

The environment mechanical properties are also relevant to cancer (Paszek and Weaver 2004; Shieh, 2011). Tumor tissue is stiffer than non-diseased tissue (Netti et al., 2000; Levental et al., 2009). On substrates with stiffness similar to normal breast tissue, mammary epithelial cells display regular mammary gland duct morphology. With increasing substrate stiffness, colonies increase in size and show disorganized morphology, similar to that of a breast cancer tumor (Paszek et al., 2005). Because metastasis and tumor invasion are key in cancer progression, much of the investigation on substrate stiffness in cancer focused on cell motility. For example, glioma cells spread and migrate rapidly on stiff substrates but were rounded and failed to migrate on softer substrates. (Ulrich et al., 2009). Other studies suggest that certain cancerous cells have altered response to substrate stiffness from normal cells (Wang et al., 2000a; Tzvetkova-Chevolleau et al., 2008). Normal 3T3 fibroblasts established synchronized traction forces at the leading lamellopodia, resulting in directed cell migration. When the fibroblasts were transformed using H-ras, traction forces appeared in various areas of the cell body in a disorganized fashion (Munevar et al., 2001). Transformed fibroblast migration speed covered a broad range and was more strongly affected by substrate stiffness, suggesting a lack of regulation (Tzvetkova-Chevolleau et al., 2008). Cancer cells can also change matrix mechanics via lysyl oxidase (LOX). Hypoxic primary tumor cells secrete LOX, which then crosslinks collagen to form a premetastatic niche (Erler et al., 2009). Areas of increased matrix stiffness due to LOX preconditioning showed increased tumor growth and invasion (Levental et al., 2009).

In diabetes, both extracellular matrix mechanics and biochemistry are altered. Elevated blood glucose levels lead to collagen glycation and crosslinking, which increases matrix stiffness. In addition, hyperglycemia causes cells to produce and secrete different amounts of matrix proteins, alters matrix protein breakdown, and may change integrin binding sites on existing proteins such as collagen (Di Lullo et al., 2002; Tsilibary, 2003). We recently showed that endothelial cells on glycated collagen did not change morphology or align their actin cytoskeleton in response to shear stress and cyclic strain, likely due to decreased focal adhesion kinase activation (Fig. 5.4) (Figueroa et al., 2011; Kemeny et al., 2011). We further showed that endothelial cell adhesion strength increased on glycated compared to native collagen, likely because cells switch from adhering with the normal collagen $\alpha_2\beta_1$ integrin to the vitronectin $\alpha_v\beta_3$ integrin (Kemeny et al., 2013). When substrate stiffness and glycated collagen were studied together, endothelial cell spread area increased with substrate stiffness on both native and glycated collagen (Figs. 5.4 and 5.5). However, cell area on native collagen increased more than 600% as gel stiffness increased from 0.3 to 30 kPa, while cell area on glycated collagen coated gels increased only slightly more than 100%. These data suggest that endothelial cell response to substrate stiffness is altered in disease conditions as cells are unable to properly spread in response to substrate stiffness on glycated compared to native collagen coated substrates. Additional studies are needed to clarify mechanisms by which changes in matrix ligand and substrate stiffness interact in varied diseases.

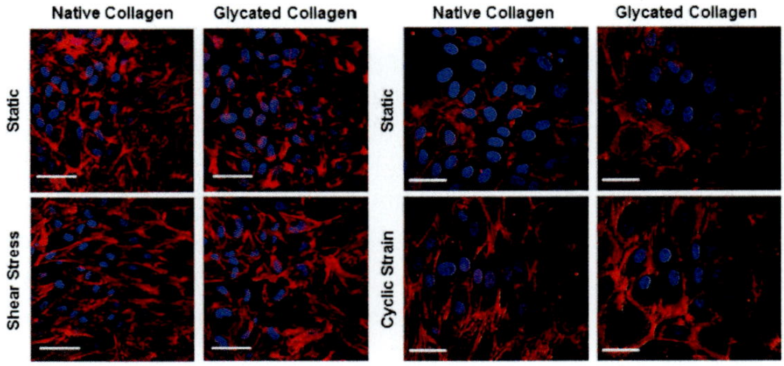

Figure 5.4 Porcine aortic endothelial cells exposed to either 20 dynes/cm^2 shear stress for 24 h (left) or 10% cyclic strain at 0.5 Hz for 6 h (right) showed decreased actin fiber alignment on glycated vs. native collagen. Cells were labeled for actin (rhodamine phalloidin, red) and nuclei (bisbenzimide, blue). Scale bar = 50 μm.

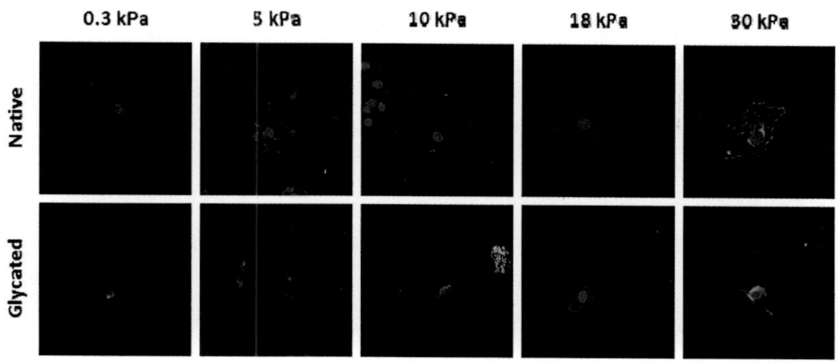

Figure 5.5 Porcine aortic endothelial cells showed larger cell spread area as substrate stiffness increased from soft (300 Pa) to physiological stiffness (5 kPa, 10 kPa) to stiff (18 kPa, 30 kPa) gels when polyacrylamide gels were coated with native but not glycated collagen. More defined actin stress fibers and focal adhesions were also observed on the stiffer collagen coated substrates. Cells were labeled for actin (rhodamine phalloidin, red), vinculin (monoclonal antibody, green) and nuclei (bisbenzimide, blue). Scale bar = 20 μm.

5.9 Cell–Substrate Mechanics: Conclusions

Cell-substrate mechanical and biochemical interactions are critical to whole tissue structure-function relationships. Continued study of the integrated effects of substrate stiffness and ligand properties will improve our understanding of the fundamental biology and physiology of native tissues. Detailed knowledge of cell-substrate interactions is further required to engineer new tissue and disease therapies. Tissue engineering scaffolds can be designed with the appropriate mechanical and ligand properties to improve tissue growth (for example by enhancing cell migration into the scaffold) and to enhance tissue function (for example by increased cardiac myocyte force production). Pharmaceutical therapies may further be able to alter extracellular matrix properties within diseased tissue and thereby stabilize atherosclerotic plaques or decrease cancer growth and metastasis. As new, more sophisticated methods are developed to study and modulate cell-substrate interactions in vitro and in vivo, we can begin to understand and treat diseases from a whole tissue perspective.

Problems

5.1 List three benefits and three drawbacks of using polyacrylamide gels to control substrate stiffness *in vitro*.

5.2 Substrate stiffness increases both cell area and cell stiffness, yet these two factors are not independent, since cell stiffness increases with cell area. Design an experiment to test whether cell stiffness increases with substrate stiffness independent of cell spread area.

5.3 The effect of matrix ligand density on cell motility changes based on substrate stiffness. Explain how this could occur.

5.4 Cell-generated traction forces are essential to their ability to migrate. Describe a disease in which it is important to understand the integrated effects of tissue stiffness and ligand density on traction forces during cell migration. How could you modulate tissue stiffness, ligand density, or both *in vivo* to treat the disease?

5.5 Identify a common mechanism for increased tissue stiffness in cardiovascular disease, cancer, and diabetes. How could this mechanism initially be adaptive?

Chapter 6

Axonal Transport and Neuromechanics

Sameer B. Shah,[a,b*] Joshua Chetta,[a] and Brian G. Bober[b]

[a]*Fischell Department of Bioengineering, University of Maryland, College Park, USA*
[b]*Departments of Orthopaedics and Bioengineering, University of California, San Diego, La Jolla, CA, USA*

sbshah@ucsd.edu

6.1 Introduction

The neuron is a cell with a distinct polarity and geometry that communicates with other excitable cells through electrochemical signaling pathways. During development, the neuron extends several processes from its cell body. One such process differentiates into an axon, a long, narrow structure that advances by means of its growth cone, a highly motile path-finding structure at its tip. The axon ultimately establishes contact with a targeted neuron in the central nervous system or a muscle fiber in the peripheral nervous system (Fig. 6.1a). At this point, the free growth cone transitions to a synaptic connection. Although synaptic termini are less motile than growth cones, they continue to dynamically respond, both biologically and structurally, to a variety of cues, including electrical activity, target movement, and the local chemical environment.

Molecular and Cellular Biomechanics
Edited by Bradley E. Layton

ISBN 978-981-4316-83-5 (Hardcover), 978-981-4613-44-6 (eBook)
www.panstanford.com

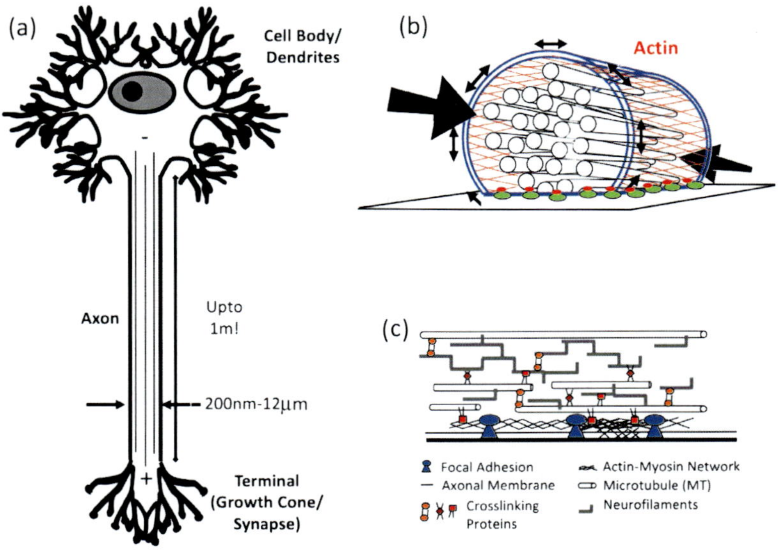

Figure 6.1 (a) Schematic of a neuron. Microtubules are oriented with their plus end projecting outward from the cell body. (b) Schematic of cytoskeletal organization within an axon. Microtubules populate the core of the axon, while actin is primarily localized to the periphery. Axons interact with their substrate through adhesion complexes. (c) Schematic of structural connectivity within an axon. Cytoskeletal elements may be cross-linked by rigid connectors or by dynamic motor proteins. Modified from Chetta et al. (2010). Reproduced with permission from Wiley-Liss, Inc.

Axonal outgrowth is accompanied by a dramatic increase in cellular volume and surface area. In fact, single axons may be up to 1 m long in humans, with volumes up to, and in excess of, 1000 times that of the supporting cell body. This places a tremendous infrastructural and metabolic demand on the cell, both during development, to enable growth, and after maturity, to maintain homeostasis as well as respond to environmental cues. This demand is met in part through the bidirectional delivery of a variety of synaptic, metabolic, and structural proteins via molecular motor-mediated axonal transport upon a network of microtubule and actin filaments. The scope of the transport challenge is astounding; transported cargoes vary in size as well as function, ranging from vesicles to mitochondria to signaling complexes to entire segments of the cytoskeleton. Adding to this complexity, axonal diameters may be as narrow as 200 nm, a dimension smaller than that of several transported cargoes themselves!

Neurons must also overcome a biomechanical challenge. During outgrowth, and also in mature neurons, stability and instability in neuronal morphology is dictated by the delicate balance of forces resulting from interactions among the multiple force-generating and load-bearing components of the axon. Elements of the cytoskeleton as well as associated motor proteins, in addition to facilitating transport, are major contributors to such biomechanical roles.

The careful organization of structural elements within the axon provides a unique and powerful model system in which interactions between structural elements may be investigated in the context of their transport as well as neuronal mechanics. Findings in neurons can subsequently be extended to elucidate general mechanisms guiding transport and biomechanics in cells with a more complex geometry. In this chapter, Section 6.2 will introduce the reader to the organization of major components of the structural framework and transport machinery within the neuron, Section 6.3 will expand upon axonal transport of the cytoskeleton, and Section 6.4 will summarize cytoskeletal contributions to neuronal cell mechanics.

6.2 Structural Organization within the Neuron

Structural networks within the neuron may be represented as a dynamic array of filaments interconnected by various cross-linking elements. This network is coupled to a substrate through adhesive protein complexes (Fig. 6.1b,c). Filaments include members of the microtubule, microfilament (actin), and intermediate filament families. Cross-linking elements include both rigid connectors and dynamic motor proteins, the latter of which rely on the hydrolysis of ATP to fuel their movement along a specified track. Considerable biochemical, cellular biological, and biophysical evidence has elucidated details of how these components are organized within the neuron. General organization of actin, microtubules, and neurofilaments within the neuron is treated in this section. Implications for such organization for transport and mechanics are treated in greater detail in Sections 6.3 and 6.4.

6.2.1 Microtubules, Kinesin, and Dynein

Microtubules are the largest of the three categories of cytoskeletal filaments and play a variety of roles within cells. These include contributing to cell structure, providing a track for protein transport, and regulating intracellular signaling. Microtubules are tubular structures 24–26 nm in diameter and are composed of dimerized α- and β-tubulin subunits. They polymerize in a concentration-dependent process through hydrolysis of a GTP molecule bound to tubulin monomers. In the axon, they exhibit a characteristic polarity, with their plus-ends, or rapidly polymerizing ends, extending away from the cell body into the axon. Structurally, individual microtubules are rigid, and typically do not branch. A variety of microtubule-associated proteins (MAPs) exist to control the stability, polymerization and polarity of microtubules as well as their interactions with motor proteins.

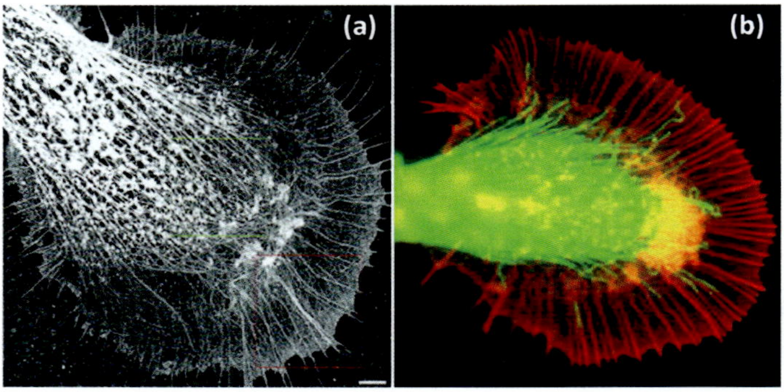

Figure 6.2 (a) Rotary shadow electron micrograph of the cytoskeleton in the Aplysia bag cell growth cone and distal axon. Bar, 3.5 μm. From Schaefer et al. (2002). (b) Fluorescence image of microtubules (green) and actin (red) in the distal axon and growth cone of an Aplysia bag cell. Modified from Munnamalai and Suter (2009)/Suter Lab Website http://suterlab.bio.purdue.edu/research.html.

Spatially, microtubules in the neuron are oriented longitudinally within the core of the axon, with their plus ends extending away from the cell body (Fig. 6.1a,b). In dendrites, microtubules exist with mixed polarity. Post-translational modification events such as detyrosination or acetylation result in a more stable population of polymerized microtubules in the mature shaft of the axon, while in the vicinity of the growth cone, microtubules polymerize and depolymerize more dynamically (Baas and Black, 1990) (Fig. 6.2). Interactions of microtubules with actin, both laterally along the axon as well as longitudinally, within the growth cone, will be discussed in Section 6.2.4.

Microtubules also serve as tracks to which the molecular motors kinesin and dynein bind. The majority of kinesins are oriented to move toward the plus end of microtubules, while dynein and a few minus-end oriented kinesins move toward the minus end of microtubules (Fig. 6.3). Generally, this motor polarity results in the movement of vesicular cargoes or other organelles in the direction of motor movement (Hirokawa and Takemura, 2004).

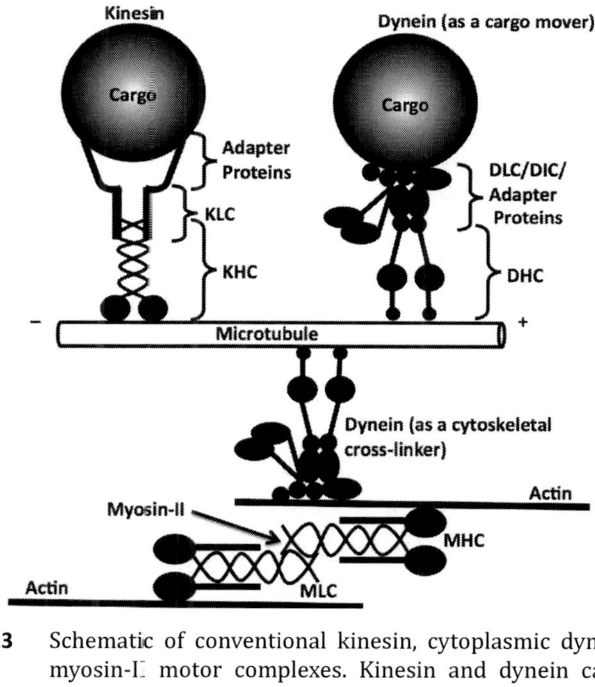

Figure 6.3 Schematic of conventional kinesin, cytoplasmic dynein, and myosin-II motor complexes. Kinesin and dynein can move cargoes along microtubules. Dynein is believed to cross-link microtubules to actin scaffolds, enabling anterograde transport of microtubules themselves. Contractility of the actin cytoskeleton is enabled by cross-linking with non-motile myosin-II. KHC: kinesin heavy chain; KLC: kinesin light chain; DHC: dynein heavy chain; DLC: dynein light chain; DIC: dynein intermediate chain; MHC: myosin heavy chain; MLC: myosin light chain.

However, recent research also indicates that microtubule-based motors also cross-link microtubules to each other or to other cytoskeletal structures (He et al., 2005; Myers and Baas, 2007), thereby influencing the movement and stabilization of cytoskeletal cargoes. MAPs, in addition to their impact on microtubule structure and stability, also regulate the interaction of microtubules with motor proteins.

6.2.2 Actin and Myosin

Actin filaments have the smallest diameter of the three classes of cytoskeletal filaments, and, like microtubules, participate in a variety of cellular activities, including motility, regulation of cellular geometry, contractility, organelle transport, and intracellular signaling. Filamentous actin (F-actin) consists of helical rods 6 to 8 nm in diameter and is composed of monomeric globular actin (G-actin) subunits. Individual filaments are polarized, with monomers added to the barbed (+) end in a concentration-dependent process through hydrolysis of a bound nucleotide, in this case ATP. Depolymerization occurs at the pointed (-) end. Unlike microtubules, actin filaments exhibit considerable flexibility and possess the capacity for complex branching. A variety of actin-binding proteins play roles in regulating functions guiding the architecture of filamentous actin networks and actin-associated signaling, including polymerization and depolymerization, nucleation and branching, capping and severing, bundling and cross-linking, and phosphorylation and dephosphorylation (Letourneau, 2009). It is the diversity of control of its polymerization, depolymerization, and bundling that contributes to the variety of functional roles for actin. Multiple types of motor proteins within the myosin family also bind and move along actin through mechanisms associated with ATP hydrolysis. Among these myosins, non-muscle myosin II serves to cross-link or bundle actin filaments. Myosin II traditionally exists as a homodimer, and through ATP-dependent reorientation of its head structure, initiates contractile activity by pulling parallel actin filaments toward each other. Actin also serves as a track for short-distance cargo transport via unconventional myosins such as myosin V, VI, or X (Bridgman, 2009). It should be noted that there is a directional bias to myosin-based motility;

all myosins move toward the barbed end of actin filaments, except for myosin VI, which moves toward the pointed end.

Actin and myosin exist in different configurations and play different roles within the growth cone and along the axon. At the leading edge of the growth cone, actin filaments and associated myosin motors are aligned with their barbed ends oriented primarily toward the tip of the growth cone (Fig. 6.2a,b). The orientation of actin and myosin is increasingly mixed in polarity in more proximal regions of the growth cone (Lewis and Bridgman, 1992). This architecture in combination with the carefully regulated polymerization and depolymerization of filaments in the growth cone is believed to enable axonal outgrowth and the production of axonal tension, as detailed in 6.3. Within the cylindrical axon, biochemical evidence indicates that actin exists as both G-actin and F-actin (Morris and Lasek, 1984; Mills et al., 1996); the latter localizes primarily to the cortex underlying the plasma membrane (Fig. 6.1b), and is cross-linked into an extensive network by myosin II (Figs. 6.2 and 6.3). This results in contractile compression, both radially and longitudinally, as described in more detail in Section 6.4.

6.2.3 Neurofilaments

Neurofilaments (NF) are the most abundant cytoskeletal proteins of myelinated axons. Neurofilaments are type IV members of the intermediate filament family (Fuchs et al., 1994), and their diameter of 10 nm is, as their categorization suggests, in between that of actin filaments and microtubules. They are obligate heteropolymers of neurofilament light (NF-L) protein with medium (NF-M) or heavy (NF-H) proteins (Ching and Liem, 1993; Lee and Cleveland, 1996). Much of the information regarding neurofilament organization, transport, and function has come through the study of mice lacking one or several neurofilament subunits or subunit domains. Mice lacking the obligate NF-L neurofilament subunit also display dramatic declines in the expression level (>20-fold) of NF-M and NF-H protein in both the central and peripheral nervous systems (Zhu et al., 1997). Consequently, there is a scarcity of intact neurofilaments in NF-L-null axons. Mice expressing defective NF-M display motor axons of smaller caliber, while mice lacking functional NF-H subunits display axons of normal caliber, but a reduced rate of radial expansion. Mice expressing truncations in both NF-M and NF-H, display bowed axonal neurofilaments devoid of cross-bridges and a disorganized neurofilament array (Garcia et al., 2003). Interestingly, recent evidence suggests that a fourth intermediate filament, α-internexin, also integrates into the mature neurofilament network. While its absence did not alter the expression or distribution of NF-L, NF-M, or NF-H proteins, the absence of NF-M and NF-H dramatically decreased the expression of α-internexin (Jacomy et al., 1999; Levavasseur et al., 1999; Munnamalai and Suter, 2009; Barry et al., 2010; Yuan et al., 2012).

Unlike other intermediate filament proteins, neurofilament heteropolymers have C-terminal side arms that project from the core of the classic 10 nm filament (Hirokawa et al., 1984). These side arms contain a repeated lysine-serine-proline (KSP) amino acid motif. Phosphorylation of serines within these repeats appears to stabilize neurofilaments, while dephosphorylation serves to increase neurofilament mobility and possibly promote degradation (Chan et al., 2003). Spatially, neurofilaments are interspersed with microtubules in the core of the axons, outnumbering their counterparts 5- to 10-fold (Lee and Cleveland, 1996). Electron microscopy reveals that axonal neurofilaments and microtubules exist in a flexible filamentous network (Fig. 6.4). In the growth cone, consistent with their increased dephosphorylation, neurofilaments are more dynamic and less integrated into the cytoskeleton (Chan et al., 2003; Uchida and Brown, 2004).

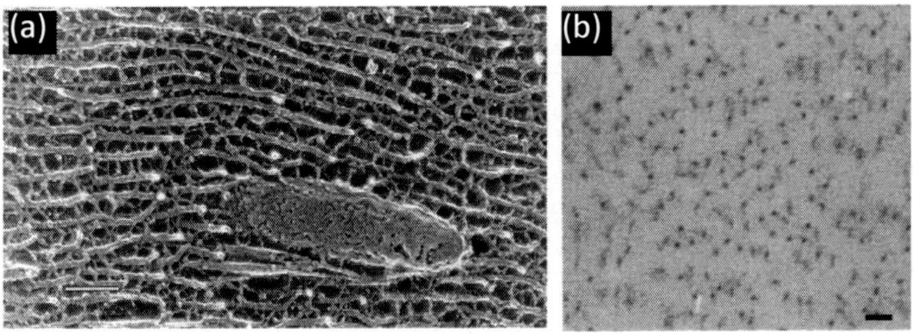

Figure 6.4 (a) Neurofilaments within an axon revealed by quick-freeze and deep-etch method. Neurofilaments are interconnected by side arm domains. Bar, 100 nm (Hirokawa and Takeda, 1998). (b) Cross-sectional electron micrographs of neurofilaments in motor axons (Rao et al., 2003).

6.2.4 Cytoskeletal Connectivity within the Axon

Multiple cross-linking proteins are responsible for creating an interconnected cytoskeleton. These may be either static load-bearing cross-links or dynamic force-generating links such as molecular motors. The complex combination of rigid and dynamic cross-linking provides a stable framework, but also one that is likely to allow axonal remodeling at a variety of length and time scales.

Mechanical connectivity between the external environment and the cell occurs via sites of focal adhesion, which are located at the cell body, growth cone, synapse, and occasionally along the axon. Extracellularly, adhesion complexes generally bind to a substrate through integrin proteins, which are then coupled to an intricate cluster of intracellular structural proteins, including talin, vinculin, paxillin, and α-actinin. Adhesion complexes serve as anchorage points for the actin–myosin network in the neuronal cortex (Burridge et al., 1987; Burridge and Fath, 1989) and are capable of modulating a variety of signaling pathways. Such pathways may act locally, such as via the recruitment and polymerization of actin or an increase in myosin activity (Ridley and Hall, 1992; Hildebrand et al., 1996). Given these capabilities, adhesion sites are likely to play a major role in the regulation of a neuron's mechanical and biological response to external loads.

At the growth cone, actin and microtubule filaments strongly influence each other. Microtubule dynamics in the growth cone may be in part modulated by the retrograde flow of actin filaments away from the tip of the growth cone and the formation of arcs of actin perpendicular to the direction of outgrowth. These processes restrict the entry of microtubules into the distal growth cone (Schaefer et al., 2002) (Fig. 6.2). In contrast, where microtubules are allowed to enter the growth cone, actin dynamics are stimulated (Bershadsky et al., 1990; Waterman-Storer and Salmon, 1999), which may then, through a positive feedback loop, enable microtubules to further polymerize and stabilize. Stable microtubules are likely to be important both for recruiting motor proteins (Nakata and Yorifuji, 1999; Konishi and Setou, 2009; Stiess et al., 2010) as well as promoting structural integrity and morphological stability.

Along the axon, actin–myosin networks in the axonal periphery are coupled to microtubules and neurofilaments within the axonal core. Actin associates with neuronal microtubules through dynein motors (Pfister, 1999; Ahmad et al., 2000; Hasaka et al., 2004; Baas et al., 2006) and through a variety of cross-linking proteins such as doublecortin, MACF, MAP1A/B, shortstop, or pod-1 (the latter two in *Drosophila*) (Fujii et al., 1993; Leung et al., 1999; Rothenberg et al., 2003; Tsukada et al., 2005; Sanchez-Soriano et al., 2009; Tint et al., 2009; Applewhite et al., 2010). Additionally, the actin cytoskeletal network interacts with neurofilaments, directly or through spectrin and myosin (Frappier et al., 1991; Hao et al., 1997; Macioce et al., 1999; Rao et al., 2002a; Jung et al., 2004). Within the core of the axon, microtubules associate with other microtubules and with neurofilaments through dynein and kinesin motors (Hirokawa, 1982; Ahmad et al., 1998; Pfister, 1999; Shea, 2000; He et al., 2005; Motil et al., 2006; Myers and Baas, 2007; Nadar et al., 2008), while neurofilaments form cross-bridges with each other through extended sidearm domains (Hirokawa, 1982; Rao et al., 2002a; Garcia et al., 2003). Side arms appear between adjacent neurofilaments and between neurofilaments and microtubules within the axon (Hirokawa, 1982; Rao et al., 2002a; Garcia et al., 2003).

6.3 Axonal Transport of the Cytoskeleton

6.3.1 Fast vs. Slow Transport

Prior to integration into a structural network, subunits of the cytoskeleton must typically be delivered in proximity to their eventual destination. Active (motor-mediated) transport is generally categorized as fast or slow based on the average velocities of a protein or cargo of interest. Such velocities were initially measured by injecting radioactive amino acids into the cell bodies of neurons such as the optic nerve and sciatic nerve, and tracking their localization at various time points following injection. Fast cargoes move an average of 20–400 mm/day (1–10 μm/sec), while slow cargoes move an average of 0.2–7 mm/day (0.01–0.1 μm/sec). Slowly transported cargoes, which include cytoskeletal and related proteins, are further subdivided into slow transport component-a and component-b, the latter being slightly faster (2–7 mm/day) than the former (0.2–2 mm/day).

Live, high-resolution fluorescence and bright-field imaging of selected cargoes in vivo has shed some light on differences between these different classes of cargo, and mechanisms underlying their transport, e.g., (Brady and Lasek, 1982; Wang and Brown, 2001). Fast and slow average velocities might result not from constant cargo motion at fast and slow speeds, but rather, via bursts of high velocities punctuated by pauses of varying duration. The fact that instantaneous velocities are identical for slow and fast cargoes are consistent with an unexpected hypothesis that the molecular motors responsible for both speeds of transport are likely to be one and the same (Ahmad et al., 1998; Wang et al., 2000b; Wang and Brown, 2001; Wang and Brown, 2002; He et al., 2005; Uchida et al., 2009). On the other hand, these discrepancies between average cargo velocity and single-motor or instantaneous velocity suggest that the motion of motor-cargo complexes requires careful regulation of the motors and their associated proteins and cargoes.

Several excellent review papers detail mechanisms of transport and the identity of various classes of motor proteins and cargoes, as well as the relevance of transport to development and disease (Hirokawa and Takemura, 2004; Chevalier-Larsen and Holzbaur, 2006; Duncan and Goldstein, 2006). It should be noted that the model systems used to examine neuronal transport vary considerably in species, identity, and stage of development. Such variability can heavily influence the interpretation of results. In this chapter, we focus primarily on broader principles governing the transport of cytoskeletal elements.

6.3.2 Microtubule Transport

Radiolabeling experiments revealed that tubulin is transported primarily in the slow component-a pool as a slowly migrating wave, though occasionally it moves with actin in slow component-b. The prevalence of full-length microtubules within the axon in combination with these experiments, which describe the average transport rate of populations of proteins rather than the instantaneous rates of individual cargoes, led to the hypothesis that microtubules themselves were continuously transported along the axon at a slow rate. Surprisingly, though, FRAP (fluorescence recovery after photobleaching) experiments in which a region of an axon expressing fluorescent tubulin was photobleached and monitored over time revealed that, in fact, axonal microtubules were largely stationary (Okabe and Hirokawa, 1992; Wang and Brown, 2002).

The paradox between a slowly moving radioactive front observed in metabolic studies with the stationary microtubule scaffold indicated by photobleaching studies was at least partially resolved through two observations. The first was the identification of populations of microtubules of varying length, ranging from a few to hundreds of microns, which were interdigitated along the length of the axon (Burton, 1987). The second was the observation through higher-resolution photobleaching assays, which revealed that individual fluorescently labeled microtubules were indeed motile at instantaneous velocities consistent with fast transport, but paused frequently and for long durations (Wang and Brown, 2002). Intriguingly, primarily shorter microtubules appear to be more mobile (Vallee and Bloom, 1991; Yu and Baas, 1994; Ahmad et al., 1998). These shorter microtubules are also believed to be of central importance to axonal growth and branching (Yu and Baas, 1994; Myers et al., 2006), and so may be reasonable candidates for influencing remodeling within the axon. A complementary perspective on the bulk transport of microtubules was presented through the use of stationary mitochondria as an indirect marker for microtubules (Miller and Sheetz, 2006). In this case, microtubules were described to move en masse as a coherent, well-connected network, through axonal tension initiated by an extending growth cone. While these data collectively provided compelling evidence that microtubules could be moved as polymers, an unresolved question continues to be the specific regions of the cell where tubulin monomers add onto shorter filaments. Experiments involving the injection of fluorescent tubulin into cell bodies suggest the incorporation of newly synthesized cell body-derived microtubules in newly synthesized regions of the axon, i.e., the growth cone (Slaughter et al., 1997). On the other hand, the recovery of fluorescent tubulin in bleached axons suggests axonal incorporation as well (Okabe and Hirokawa, 1992).

The proteins responsible for microtubule transport were elucidated through a combination of chemical and genetic perturbation. Anterograde transport of microtubules is believed to result in large part, counterintuitively, through the activity of minus-end directed microtubule based motors, including dynein and atypical mitotic motors in the kinesin-14 family. Minus-end directed motors typically move vesicular cargoes retrogradely. However, if the "cargo" offers sufficient resistance, the microtubule track moves anterogradely, in a manner analogous to the moving track of a treadmill. In the case of dynein, there is considerable evidence that an actin network is likely to serve as this rigid cargo (Pfister, 1999; Ahmad et al., 2000; Hasaka et al., 2004; Baas et al., 2006). This mechanism was confirmed by the disruption of anterograde transport by both depolymerizing the actin network chemically and disrupting the activity of dynein genetically. In the case of kinesin-associated motors, other microtubules, presumably the longer ones that are part of an extensive network, serve as the rigid cargo. Genetic disruption of these kinesins resulted in the inability of neuronal cells to extend neurites, an indirect indicator of reduced anterograde transport of microtubules. Motors responsible for retrograde microtubule transport remain to be confirmed, though kinesin-5 (Eg5), which forms a homotetramer that also cross-links microtubules, has been posited as a promising candidate (Falnikar et al., 2011).

6.3.3 Actin Transport

The transport of actin is perhaps the least well characterized of the three filament types. Early radiolabeling experiments indicated that actin is transported at rates associated with slow component-b (Black and Lasek, 1979; Willard et al., 1979), though other studies report that some actin may also travel with tubulin in both slow component-a and -b (McQuarrie et al., 1986; Tashiro and Komiya, 1989; Tashiro and Komiya, 1991). Mechanisms regulating details of actin transport remain unclear, though, in large part due to the complexity, heterogeneity, and ubiquity of actin throughout the neuron.

Early reports posited that like neurofilaments, actin also moved as a polymer "sliding" along the axon. However, radiolabeling experiments combined with biochemical evidence from DNAse I affinity chromatography assays, one- and two-dimensional gel electrophoresis, and peptide mapping suggest that much of the transported actin is soluble (Morris and Lasek, 1984; Mills et al., 1996). Although the relative contributions of monomers, oligomers, or short actin filaments to the soluble fraction is still unknown, these smaller units frequently exchange with a less mobile filamentous form along the axon or within the growth cone (Okabe and Hirokawa, 1990). Specific mechanisms regulating this exchange are not known, though actin-associated proteins such as actin depolymerizing factor, cofilin, and profilin have been reported to travel in a complex with actin (Maekawa et al., 1984; Sun et al., 1995; Mills et al., 1996; Bamburg, 1999), and therefore would be well positioned to modulate actin exchange.

Higher-resolution live imaging of fluorescent actin fusion proteins and immunolabeling also reveal actin within the axon to exist as a diffuse filamentous network superposed with puncta of actin. Interestingly though, axonal actin is more mobile than originally thought. Actin is a key component of axonal filopodia, protrusions which extend and retract and which may ultimately represent precursors to axonal branches (Ketschek and Gallo, 2010). In addition, actin has also been observed to move as a wave, associated with both central and peripheral neuronal outgrowth (Ruthel and Banker, 1998; Ruthel and Banker, 1999; Flynn et al., 2009). Waves originate at cell bodies and also travel along neurites to their tips at rates typical of slow transport component-b. These highly motile structures appear similar to growth cones, and provide a relatively large bolus of material delivery to the neuronal terminal, presumably to meet the demands of a growing cell. Finally, our own observations reveal an additional population of filamentous actin puncta along the axon, which move bidirectionally, but for shorter distances. The function of these puncta is not known, though they may be candidates to incorporate into the cytoskeleton, both along the axon and at the growth cone. The motor protein(s) responsible for actin transport have not yet been formally identified, though a myosin-like protein has been observed to be transported with actin waves (Willard et al., 1979), and unpublished data (Chetta and Shah, unpublished) indicates a role for both actin/myosin and microtubule-based transport of actin.

While the basic principles of actin transport, as described above, are likely to be conserved in neurons of considerable variety, reported bulk rates of axonal actin transport vary from species to species, between central and peripheral neurons, and even between various branches of the same nerve (Mori et al., 1979; Archer et al., 1994). The logic underlying these differences is not clear, though they may result from differences in protein expression and localization, material requirements, or cytoskeletal architecture.

6.3.4 Neurofilament Transport

There are numerous parallels between microtubule transport and neurofilament transport. Radiolabeling and biochemical experiments revealed that neurofilaments are transported primarily in the slow component-a pool as a coherent wave, and exist axonally in a predominantly polymerized state. This result was consistent with the early proposal that cytoskeletal polymers slide along other cytoskeletal polymers within in axons (Black and Lasek, 1980). Like microtubules, though ultimately neurofilaments form a dense, interconnected network in mature regions of the axon (Hirokawa, 1982; Leterrier and Eyer, 1987; Lee and Cleveland, 1996; Leterrier et al., 1996; Garcia et al., 2003; Barry et al., 2007; Rammensee et al., 2007), those filaments observed to translocate along axons appear to be shorter in length (Yabe et al., 1999; Shea, 2000; Wang and Brown, 2001). Also like microtubules, the slow net rate of neurofilament transport was not a result of a slow motor, but rather, fast transport punctuated with long pauses (Wang and Brown, 2001). Consistent with this observation, the motor proteins responsible for neurofilament transport are kinesin 1 and dynein, for anterograde and retrograde transport, respectively (Yabe et al., 1999; Wagner et al., 2003; Xia et al., 2003; Barry et al., 2007).

The phosphorylation state of NF-H and NF-M side arms appears to be critical for neurofilament transport and distribution through the axon. This relevance has been inferred through the examination of myelinated neurons, which undergo two distinctive phases of growth. During development, peripheral neurons extend axons to prospective targets. After stable synapse formation, a second phase of radial axonal growth is initiated, in which axonal neurofilament density increases and axonal diameter is increased up to tenfold (Cleveland, 1996; Williamson et al., 1996). Myelination, which is required for neurofilament phosphorylation, is essential for this radial growth phase (de Waegh et al., 1992; Garcia et al., 2009). The decreased phosphorylation of neurofilaments and the reduced accumulation of neurofilaments following demyelination suggests that phosphorylation of NFs negatively regulates the transport, and thus accumulation, of neurofilaments (Barry et al., 2007). This is consistent with motility and biochemical data suggesting that phosphorylation of neurofilaments has a strong influence on neurofilament transport and interactions with motor proteins (Yabe et al., 1999; Ackerley et al., 2003; Jung et al., 2005). The identity of the specific neurofilament subunit and the presence of side arms are also likely to influence transport. Analysis of mice expressing neurofilaments lacking NF-M side arms, but not NF-H side arms, reveals impaired radial growth. On the other hand, transport rates of either subunit were not affected by truncation of one or the other subunit, suggesting some compensatory phosphorylation of the other subunit (Rao et al., 2002b; Rao et al., 2003). Assessment of transport in mice lacking sidearms from both subunits will provide further insight into the roles of sidearms and their phosphorylation on transport (Barry et al., 2007). It should also be noted that though the rates of neurofilament transport in certain mutants were unaffected, the reduced caliber or rate of radial expansion of axons in NF-M or NF-H mutants necessarily results from a reduced flux at some stage of development. Therefore, aside from mechanisms regulating the movement of individual filaments, regulation of the frequency of neurofilament entry into the axon and pause duration within the axon are an interesting and unexplored aspect of neurofilament transport.

6.3.5 Mechanical Determinants of Transport

A conceptual model defining mechanical influences on transport may be subdivided into four components: (i) the inherent load-generating and load-bearing properties of specific motor-proteins; (ii) the configuration of cooperative and competitive motors on a given cargo at any given time; (iii) cargo geometry; (iv) forces opposing transport, including drag forces on a single cargo, or mechanical properties of a network into which a cargo is cross-linked. We briefly address each of these components, with the final component providing a natural bridge from cytoskeletal transport to neuronal mechanics.

6.3.6 Inherent Motor Capabilities

A number of elegant in vitro experiments have elucidated the biophysical capabilities of myosin, kinesin, and dynein motors, including their movement upon actin and microtubules. (The discussion below synthesizes results obtained from several excellent references (Howard et al., 1989; Vale et al., 1992; Kuo and Sheetz, 1993; Coppin et al., 1996; Kawaguchi and Ishiwata, 2000; Mallik et al., 2004; Mallik and Gross, 2004; Mallik et al., 2005; Gennerich et al., 2007).) Each family of motor proteins has many members, resulting in tremendous functional diversity. Unless noted otherwise, we will restrict our discussion to conventional myosin (-II), conventional kinesin (-1), and cytoplasmic dynein (Fig. 6.3). Structurally, the functional core of myosins, kinesins, and dyneins are heavy chains, which contain globular head domains that bind ATP and enable its hydrolysis. The energetics of these reactions serve to regulate motor attachment to their respective filament tracks and movement. Under physiological conditions, kinesin and myosin motors have associated light chains to help with motility and cargo binding. Dynein too has several isoforms of light chains, but also a variety of intermediate chains to help regulate motility and cargo attachment. Dynein also often exists with an additional regulatory complex of proteins known as dynactin.

In its simplest sense, the function of a single motor complex can be defined based on load-bearing (force-generation) capabilities, step size, velocity, and duty ratio (proportion of time bound to a filament). Additionally, though myosins, kinesins, and dynein all use ATP hydrolysis to enable stepping, there are differences in the ATP dependence of step size and force generation. These functional differences are in large part due to structural differences as well as coordination of ATP binding, hydrolysis, and nucleotide and phosphate release within the complex of motor proteins.

For example, conventional myosin and kinesin both exist as dimers, and both require a single ATP binding and hydrolysis event to generate force. However, the two heads of kinesin move in a coordinated fashion, while myosin heads bind and detach independently. Additionally, kinesin heads stay bound for a longer fraction of the duty cycle than myosin heads. Finally, kinesin expends one ATP per step, while myosin may move several steps with one ATP. As a consequence, the motility of kinesin is quite different from myosin. Kinesin walks in a processive manner, often moving more than one hundred 8 nm steps at 1 μm/sec before falling off of a microtubule. In contrast, myosins are not processive, but cooperative. Individual myosins bind in a stochastic manner and detach rapidly, moving along actin for a few ~5–8 nm steps resulting in total displacements of 10–50 nm before detachment; consequently, multiple myosins are required to ensure that at least one motor is bound and generating force at any given time. Dynein motility has unique motility characteristics from both kinesin and myosin. In one sense, dynein is thought to be a processive motor, since it takes multiple steps in a coordinated manner similar to kinesin. On the other hand, like myosin, dynein is thought of a cooperative motor, as increased motor number reduces the likelihood of cargo detachment from a microtubule track. Unlike either kinesin or myosin, dynein heavy chain possesses multiple ATP binding sites, which are partially or fully filled with ATP molecules. Increased load results in multiple bound ATPs and increased force production, but decreased step size. In contrast, decreased load results in fewer bound ATPs and a longer step size.

6.3.7 Motor Configurations on a Cargo

While in vitro, conditions can be manipulated to isolate the activity of a single motor protein, whereas within cells, a given cargo is often bound to multiple motor proteins varying in number and identity. Given the differences in individual motor characteristics, including force-generating capabilities, directionality, processivity, and cooperativity, the specific configuration of motor proteins on a cargo can have a dramatic effect on cargo movement. This is especially true for cytoskeletal elements, which simply by virtue of their size and variety of associated motors and cross-linking proteins can be pulled, pushed, and stabilized in myriad combinations. One might imagine a scenario whereby a single cargo is bound to dynein, kinesin, and myosin (Welte et al., 1998; Gross et al., 2002); understanding the nature of motility of this cargo is a matter of great interest and complexity. Questions regarding communication between various motors on a cargo, both of the same type (e.g., one kinesin vs. multiple kinesins) or of opposing motors (e.g., kinesin vs. dynein) are still under debate. Several interesting biological and theoretical models for coordination and competition have been proposed, including the possible role of dynactin or other proteins in the communication between kinesin and dynein (Haghnia et al., 2007; Kardon and Vale, 2009; Hendricks et al., 2010), differential loading of similar motors on a cargo based on their proximity to the moving edge of the cargo (Zhang, 2011), or assorted variations of a tug-of-war and cooperativity model between motors of opposing polarity (Vale et al., 1992; Badoual et al., 2002; Muller et al., 2008).

6.3.8 Influences of the Cellular Environment

In addition to superposition or coordination of individual motor proteins, the extracellular environment and cargo geometry may also play a role in motor function, in creating and defining the forces encountered by the motor proteins. The most intuitive example of such influences is the classical force–velocity curve in skeletal muscle, where increased coupling of myosin to actin results in increased force generation, at the expense of filament sliding velocity. A similar relationship is seen for dynein, another cooperative motor, where increased load results in reduced step size (Mallik and Gross, 2004). Interestingly, both in vitro and in vivo experiments suggest that, contrary to predictions based on single motor studies, kinesin can behave cooperatively in the presence of increased loads, effected by larger cargoes or increased viscosity (Hill et al., 2004; Gagliano et al., 2010; Zhang, 2011).

In the context of moving or sliding cytoskeletal filaments, the length of a given filament is of critical importance. Filament length not only dictates filament mobility through the viscous axoplasm but also is likely to influence cross-link or motor number; shorter filaments are likely to be more mobile, as they will experience less drag, and are less likely to be interconnected into a broader scaffold. This logic is consistent with previously reported observations of microtubule and neurofilament mobility alluded to earlier in this chapter. It is therefore likely that the longest filaments are most integrated into the cellular cytoskeletal network and are therefore most likely to play the most significant role in defining the bulk mechanical properties of neurons. The next section provides additional details on neuromechanics.

6.4 Neuromechanics

Peripheral and central neurons experience mechanical loading in many scenarios. In the peripheral nervous system, nerves are loaded during development and growth, as well as during the flexion and extension of joints, which are spanned by the nerves (Topp and Boyd, 2006). Non-physiologically, nerves are loaded during injury, entrapment neuropathies (e.g., carpal tunnel syndrome), physical therapy or surgery (Bora et al., 1980a; Spiegel et al., 1993; Ikeda et al., 2000; Jou et al., 2000; Abe et al., 2002; Coppieters et al., 2002; Abe et al., 2003; Yokota et al., 2003; Abe et al., 2004; Ichimura et al., 2005; Lee et al., 2006; Li and Shi, 2006; Topp and Boyd, 2006; Gilbert et al., 2007b; Gilbert et al., 2007a). In the central nervous system, neurons are also loaded during development and growth. Additionally, neurons can accommodate physiological fluctuations in osmotic or hydrostatic (intracranial) pressure. Neurons are subject to more dramatic changes in their mechanical environment non-physiologically in response to traumatic brain injury (TBI) and disease (Bandak, 1995; Goldsmith and Plunkett, 2004; Wostyn et al., 2008). Mechanical loads encountered by neurons may be broadly categorized as compressive, tensile (stretch), or shear (or some combination of these components). The deformation in response to such loading is dependent upon the material properties of the tissue being loaded. Neurons, as typical for most biological soft tissues, are viscoelastic, indicating an immediate elastic response and a time-dependent viscous response to deformation. While shear and compressive forces are likely to be important regulators of neuronal function, we will focus the bulk of our attention on tensile loading given its relevance to neuronal development and regeneration. A brief summary of compressive and shear loading is provided for reference.

6.4.1 Tensile Loading during Normal Axonal Outgrowth

The sequential assembly of actin filaments at the leading edge (tip) of the growth cone and the myosin II-mediated disassembly of these filaments proximally results in a retrograde flow of actin that propels the growth cone forward. These actin dynamics are also coupled to signaling cascades that stabilize the terminal axon against its substrate through the activity of adhesion complexes. Upon establishing such adhesion sites, traction forces are generated to promote and sustain axonal extension.

During development, growth cone advancement places the axon under tension and requires a series of attachment (adhesion) and detachment events, which occur in coordination with cyclic changes in the stability of actin filaments and associated myosin motor proteins. Within the axon, the contractile activity of cortical actin–myosin networks creates longitudinal compressive forces, which are borne by microtubules and possibly neurofilaments.

At the cell level, axons are under tension during growth and in response to external loads. During growth, axial (longitudinal) tension in the axon is generated by actin–myosin interactions within the advancing growth cone and along the axon. This observation was first noted by Bray, who also proposed a role for the cellular cytoskeleton in the stabilization of neurons (Bray et al., 1978; Bray, 1979; Bray and Gilbert, 1981). Heidemann and colleagues subsequently elucidated interactions between actin and microtubule elements (Heidemann et al., 1985; Joshi et al., 1985; Dennerll et al., 1988). These investigators transected axons and treated cells with actin and microtubule destabilizing drugs. They then assayed whether neurite retraction (recoil), indicating tension within the neurite, occurs. The typical retraction observed following neurite transection was eliminated upon treatment with cytochalasin, a drug that disrupts actin networks (Joshi et al., 1985). On the other hand, treatment with nocodazole, which destabilizes microtubules, suppressed neuronal growth and induced rapid retraction. These results suggest that the regulation of microtubule polymerization has a strong influence on both mechanical properties of the neuron and neurite outgrowth (Heidemann et al., 1985; Dennerll et al., 1988). Collectively, these studies revealed that the compression of microtubules during normal axonal growth counteracts the tension generated within the actin network.

Experiments on motor proteins associated with actin and/or microtubules confirmed conclusions from experiments on filament proteins described in Section 6.2.1. Average axonal growth rates (~10–50 μm/h) reflect periods of extension followed by periods of pausing or retraction rather than uninterrupted elongation. Baas and colleagues examined the interplay between microtubules and actin networks in guiding such patterns of growth, by investigating roles of the microtubule-associated motor protein dynein and the actin-associated motor protein myosin II on guiding the extension and retraction of axons (Myers et al., 2006). Microtubule advancement into the growth cone was impaired following the inhibition of dynein, resulting in increased axonal retraction. Conversely, the inhibition of myosin II allowed for microtubule advancement into the growth cone and filopodia, thereby halting retraction. These data led to a model in which dynein-driven forces encourage expansion by opposing myosin-II driven forces that favor retraction. This model provides one explanation for pauses in growth cone motility, as actin-based and microtubule-based forces may occasionally counteract each other, leading to a stalemate between extension and retraction.

Miller and Sheetz (2006) provide an alternate explanation for pauses during extension in their examination of the relationship between mitochondrial movement and axonal growth. The investigators suggest that the axonal framework is not stationary, but undergoes low velocity transport that results in axonal growth through a mechanism independent of, but not necessarily in opposition to, growth cone advance. They suggest that periods of growth are followed by the addition of required cellular materials, including mitochondria, which are delivered by fast axonal transport (Lin and Forscher, 1995; Challacombe et al., 1996).

6.4.2 Imposed Tensile Loading

Qualitative observations of mechanisms underlying the generation of and response to tension in the neuron during normal growth have been extended by imposing tension externally upon individual axons with free growth cones (Fig. 6.5). Initial "towed-growth" experiments using a micropipette (Bray, 1984; Zheng et al., 1991; O'Toole et al., 2008) resulted in accelerated growth compared to unstretched axons, in a manner dependent on strain and strain-rate. Under moderate loading conditions, chick DRG axons stretched or deflected with a micropipette display viscoelastic changes in load-bearing capabilities, including stress relaxation and creep. Stiffness for these axons was estimated to be between 15–160 μdyn/μm, based on force-displacement curves obtained from calibrated glass needles (Lamoureux et al., 1992). Actin's contribution to these estimates of stiffness was quantified;

following cytochalasin treatment, these constants dropped to <2.5 μdyn/μm, or <16% of the reference value (Dennerll et al., 1988). Smith and colleagues dramatically extended micropipette studies by coaxing tracts of parallel axons into an "extreme stretch growth" regime. Through the use of a bioreactor, axons were stretched at a rate of several mm/day to lengths of over 10 cm in length (Smith et al., 2001; Pfister et al., 2003; Pfister et al., 2004; Pfister et al., 2007) without loss of electrical function.

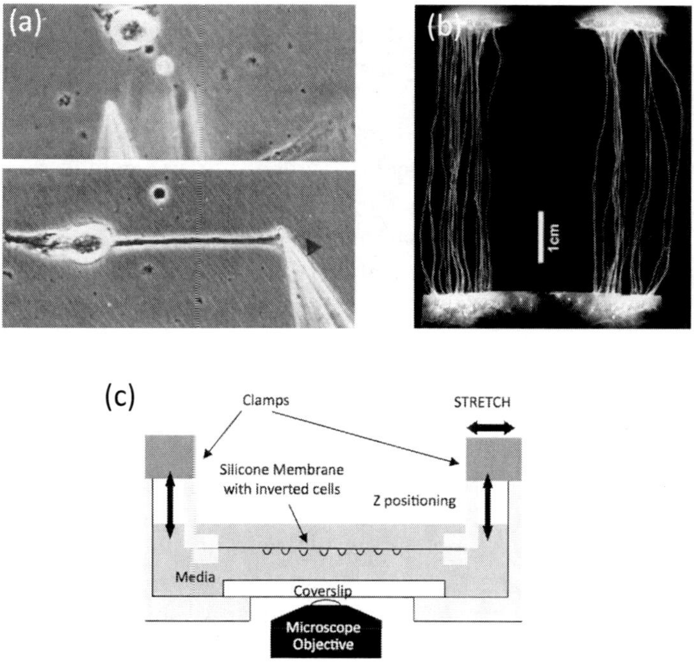

Figure 6.5 A variety of experimental approaches have provided insight into the response of neurons to tensile loading (a). (top) Neuron prior to needle attachment and (bottom) same cell as in previous frame after 96 min of experimentally applied tension with a micropipette ("towed-growth") (Zheng et al., 1991). (b) A mechanical bioreactor yields extremely long tracts of "stretch-grown" axons (Pfister et al., 2004). (c) Schematic of a device used to uniaxially load cells seeded on a flexible substrate ("substrate-strain"), which allows high-resolution imaging of neuronal response to stretch. Chetta, J., C. Kye and S. B. Shah (2010). Cytoskeletal Dynamics in Response to Tensile Loading of Mammalian Axons. *Cytoskeleton (Hoboken)*, **67** (10):650–665. For full resolution images.

Mechanisms underlying the accommodation of such dramatic growth are still being resolved. Intriguingly, such mechanisms are likely to be different from those underlying normal growth mechanisms. By using styrene microspheres to label neurite surfaces, it was determined that for stretched axons, membranous material is added along the length of the entire axon in an unpredictable manner. On the other hand, naturally growing axons tend to add material primarily at the growth cone (Bray, 1984; Zheng et al., 1991). In addition, recent work using a substrate-strain experimental model, where neurons are attached to a flexible substrate that is stretched uniaxially, suggest that the mechanical properties of a cell vary regionally (Chetta et al., 2010). These observations raise the interesting hypothesis that regions of high mechanical deformation may be coupled to the targeted transport of cellular components responsible for augmenting a cell.

Finally, while tensile loading such as that experienced during traumatic brain injury is outside of the scope of this chapter, it is worth noting that recent models of rapid axonal injury indicate microtubule breaking (Tang-Schomer et al., 2012) and calpain-mediated disintegration of the spectrin-actin cytoskeleton (Buki and Povlishock, 2006; Huh et al., 2006) as key modes of failure, again emphasizing the relevance of these cytoskeletal components to axonal integrity and mechanical function.

Phenomena at the cell level are consistent with those at the tissue level, where neurons typically do not possess a free growth cone. Peripheral nerves in situ are also under tension, and release strains of up to 11% following transection (Rydevik et al., 1990; Kwan et al., 1992; Brown et al., 1993). During physiological movement, nerves slide in their bed (excursion) and unravel fiber undulations known as bands of Fontana (Haninec, 1986; Topp and Boyd, 2006) before additional increases in nerve tension. Animal models of limb lengthening and ex vivo biomechanical models reveal accelerated nerve growth and remodeling at moderate strains (Bora et al., 1980a; Spiegel et al., 1993; Ikeda et al., 2000; Jou et al., 2000; Abe et al., 2002; Abe et al., 2003; Yokota et al., 2003; Abe et al., 2004; Ichimura et al., 2005; Lee et al., 2006; Li and Shi, 2006; Bueno and Shah, 2008). Tensile material properties of peripheral nerves reveal a wide range of values, with elastic moduli ranging 8–40 MPa, stress relaxation of up to 57%, and peak stresses of up to 11.7 MPa in response to strains of 2 to 50% (Rydevik et al., 1990; Wall et al., 1992; Pfister et al., 2004; Ichimura et al., 2005). These differences are likely to reflect differences in species, nerve branch (e.g., sciatic vs. tibial), and differences in boundary conditions (in situ vs. ex vivo; different mechanisms of clamping). Recent studies provide a potential structural basis for the nervous response to tensile loading; increasing loads induce failure first in the perineurial interface, followed by the nerve core, composed of axons and endoneurium, and, finally, the epineurial sheath (Walbeehm et al., 2004; Georgeu et al., 2005). Additional experiments are required to test whether this hierarchy also underlies stress relaxation and other viscoelastic properties of moderately loaded nerves.

Finally, in light of recent work describing the dramatic influence of substrate stiffness on the identity of cells differentiated from stem cells (Engler et al., 2006), it is worth a brief comment on the influence of substrate stiffness on neuronal growth. The properties of the substrate influence neuronal growth, but not necessarily by altering adhesive properties of the growth cone (Gundersen, 1988; Lamoureux et al., 1992; Zheng et al., 1994). Rather, elongation rate is traditionally expressed strictly as a function of axonal tension, irrespective of the substrate (Lamoureux et al., 1992). Since axonal tension is modulated by the nature of neuronal anchorage to its substrate, substrate stiffness is likely to have a strong influence on rates of elongation. Consistent with this logic, a strong negative correlation between stiffness and elongation was observed for the migration of rat neuroblastoma cells on hydrogels of varying stiffness (Gunn et al., 2005). On the other hand, evidence in hippocampal neurons and rat neuroblastoma cells (Leach et al., 2007; Previtera et al., 2010a; Previtera et al., 2010b) indicates that stiffness influences axonal branching, but not necessarily the rate of neurite elongation. These findings preclude a simple relationship between substrate stiffness, axonal tension, and axonal growth rates. In attempting to resolve some of these apparent discrepancies, it is necessary to consider that the discussed experiments have been performed on a variety of neuronal cells, including neuroblastoma cells, peripheral neurons, and central neurons from several different organisms. In light of potential structural and functional differences among these diverse neurons, conclusions from each experiment must be placed within the context of a specific model system. For example, different rates of neuronal extension and different axonal transport and adhesion patterns have been reported in a number of species (Okabe and Hirokawa, 1992; Popov et al., 1993; Futerman and Banker, 1996). Because the distribution of adhesion patterns along an axon influences the distribution of tension along the neurite, specific mechanical properties may not be common to all neurons, but are instead associated with a specific experimental model. This hypothesis is borne out in work comparing the micropipette-assisted elongation of chick peripheral and forebrain neurons. While a modulus of elasticity could be readily calculated in peripheral neurons, "no minimum threshold tension [was] required for elongation in forebrain neurons" (Chada et al., 1997).

Interestingly, the contributions of neurofilaments to neuronal biomechanics have not been directly tested. The roles of neurofilament side arms have been especially well characterized, with C-terminal truncations in mouse NF-M and/or NF-H proteins revealing their role in the regulation of axonal caliber and the rate of radial growth, respectively. Cumulatively, truncations in both NF-M and NF-H result in bowed axonal neurofilaments devoid of cross-bridges and a disorganized neurofilament array (Garcia et al., 2003). Given this structural impact, it is likely that neurofilaments will dramatically impact the mechanical properties and biomechanical response of axons.

6.4.3 Compressive and Shear Loading

The biomechanical response of individual neurons to compression and shear loading has been studied less intensively than its response to stretch. Osmotic compression of neurons results in beading along the axon and a reduced axonal diameter (Pullarkat et al., 2006; Fernandez and Pullarkat, 2010). The osmotic environment of a neuron also appears to influence its response to tensile loads. In culture medium diluted by 50%, neurites grow faster than in undiluted media through a hypothesized mechanism that results from the release of compression on microtubules in the axonal shaft (Ahmad et al., 1998). Compressive mechanical properties of neurons have been poorly defined, though cell bodies are likely to reflect similar material properties as other cells (~1 kPa) (Sirghi et al., 2008).

At the tissue level, nerve compression may result as radial contraction following stretch or due to extraneuronal pressure or indentation (Millesi et al., 1995; Topp and Boyd, 2006). Several studies report structural and functional damage to myelinated neurons in response to compressive pressures ranging from ~7–40 kPa. Circular compression of the rabbit sciatic nerves reports a transverse compressive modulus of ~70 kPa (Ju et al., 2006), 1–2 orders of magnitude smaller than the reported range of tensile moduli noted in the previous section. Few, if any, other studies have measured the deformation of peripheral nerves or structural mechanisms underlying neuronal plasticity in response to compression. However, a few studies performed on the spinal cord (e.g., Ouyang et al., 2008) suggest compressive moduli almost two orders of magnitude higher than those of peripheral neurons. Shear loading has not been studied in the peripheral nervous system. In the central nervous system, shear loading has primarily been examined in the context of traumatic brain injury (LaPlaca et al., 2005; Cullen and LaPlaca, 2006; LaPlaca et al., 2006), and in the vestibular system, in the context of fluid shear (LaPlaca et al., 2005; Cullen and LaPlaca, 2006; LaPlaca et al., 2006).

6.5 Summary and Outlook

There is considerable physiological and clinical relevance for both axonal transport and neuronal mechanics. Although the literature in each topic has gradually increased, there remains considerable room for research at both the mechanistic and the systems level. In addition, this chapter has described considerable overlap in the mechanisms responsible for cytoskeletal transport as well as the regulation of neuronal mechanics. Consequently, it is recommended that an examination of either of these themes should be interpreted taking into consideration the other.

Problems

6.1 List three types of motor proteins that exist in neurons, and the tracks on which they move.

6.2 A neuron in humans can be up to 1 m in length. In such a neuron, approximately how long would it take a vesicle to leave the cell body and reach the neuronal terminus? A microtubule?

6.3 A drug is introduced that increases neurofilament cross-linking to other neurofilaments and to microtubules. How might this affect transport?

6.4 What structural elements are responsible for bearing tensile loads in individual neurons?

6.5 What structural elements would create differences in the elastic moduli of neurons and whole nerves?

6.6 List two protective mechanisms that allow neurons or nerves to bear tensile loads.

Chapter 7

Implications for Disease: Valvular Fibrosis and the Myofibroblast

Mark C. Blaser[a,b] **and Craig A. Simmons**[a,b,c]

[a]*Institute of Biomaterials and Biomedical Engineering,*
[b]*Cardiovascular Sciences Collaborative Program, and*
[c]*Department of Mechanical and Industrial Engineering, University of Toronto, 5 King's College Road, Toronto, Ontario, M5S 3G8, Canada*

c.simmons@utoronto.ca

7.1 Introduction

With increased understanding of biomechanics at both cellular and molecular levels have emerged new insights into the etiology and pathogenesis of a wide variety of disease states. A particularly important pathology that is intimately connected with tissue, cellular, and molecular mechanics is fibrosis.

7.1.1 Fibrosis: A Disease Rooted in Mechanics

Fibrosis manifests in a wide variety of tissues and organs and represents an insidious hijacking of the body's connective tissue repair mechanisms. In healthy tissue, there exists a finely balanced

Molecular and Cellular Biomechanics
Edited by Bradley E. Layton

ISBN 978-981-4316-83-5 (Hardcover), 978-981-4613-44-6 (eBook)
www.panstanford.com

homeostasis between extracellular matrix (ECM) synthesis and degradation (Everts and Beertsen, 1996). In normal wound healing, the ECM and tissue are restored such that healthy tissue structure, composition, and function are replicated. When these homeostatic systems become deregulated, ECM excess can be secreted in the form of disorganized fibrous collagen, which devolves into stiff scar tissue (MacKenna et al., 2000; Wynn, 2007). At first, this serves an important role, as the contractile cellular components of scar tissue enable speedy closure of open wounds. Subsequently, however, continued fibrotic ECM overproduction can affect organ and tissue function, including that of skin, heart, blood vessels, kidneys, lungs, and liver. Permanent, stiffened fibrotic scar tissue alters tissue mechanical properties and inhibits natural organ function (Hinz, 2009).

Biomechanics, from the tissue to molecular levels, plays a central role in the pathological organ-level fibrosis development. The tissue stiffening inherent to scar formation establishes a positive-feedback system whereby the mechanical properties of the scar microenvironment mediate differentiation of healthy fibroblasts to a contractile, ECM-synthesizing myofibroblast (Hinz, 2009). The myofibroblast then promotes further tissue contracture and matrix protein overproduction, and the cycle begins anew. In addition, stiffened fibrotic ECM no longer shields its resident cells from external stresses, likely leading to further myofibroblast differentiation (Eastwood et al., 1998; Tomasek et al., 2002; Hinz et al., 2007). While fibrosis has been implicated in the initiation, development, and progress of a wide variety of diseases and conditions (Table 7.1), in this chapter we focus on the aortic heart valve and its propensity to undergo pathological fibrotic remodeling. The complex combination of tissue morphology, cellular composition, mechanics, and biochemical signaling present in the aortic valve make it an excellent model system for other fibrotic conditions. In some cases, the cellular mechanisms through which valve cells contribute to valve pathology are poorly defined, and so we draw on work involving other cardiovascular and fibroblastic cell types in order to expand upon concepts and models.

Table 7.1 Selected pathologies linked to fibrosis

Pathology	Description
Cancer	Tumors and their associated stroma are stiffer than normal connective tissue; myofibroblasts aid in preparing the surrounding stromal tissue for invasion by a growing tumor.
Cirrhosis	Chronic liver diseases result in the replacement of liver tissue by fibrotic scar tissue as hepatic stellate cells differentiate to myofibroblasts. Leads to blockage of portal blood, abdominal fluid retention (ascites), and impaired liver function.
Crohn's disease	Inflammatory tissue damage results in presence of both subepithelial and Cajal myofibroblasts; often requires surgical intervention.
Cystic fibrosis	Chronic genetic disorder in which a mutated ion channel results in thickened pancreatic juice, plugged pancreatic ducts and myofibroblastic differentiation of pancreatic stellate cells.
Keloid scarring	Prototypical impaired dermal wound healing; continued production of fibrotic ECM and scar tissue after wound closure.
Muscular dystrophy	Skeletal muscle-wasting disease; degenerated muscle fibers push myoblasts to differentiate to myofibroblasts, while fibrotic and adipose tissue replaces skeletal muscle.
Myocardial fibrosis	Interstitial cardiac fibroblasts proliferate, differentiate to myofibroblasts, and secrete substantial amounts of collagen, resulting in cardiac hypertrophy and/or heart failure.
Pulmonary fibrosis	Often idiopathic; pulmonary fibroblasts differentiate to myofibroblasts, stiffen the lungs and reduce compliance.
Renal disease	Resident fibroblasts and tubular epithelial cells differentiate to myofibroblasts, resulting in fibrosis and declining kidney function.

(*Continued*)

Table 7.1 (*Continued*)

Pathology	Description
Response to device implantation	Non-specific immune "Foreign Body" inflammatory response; tissue resident fibroblasts differentiate to myofibroblasts and encapsulate foreign bodies and implanted biomaterials in a thick fibrous capsule.
Valvular fibrosis	Quiescent valvular interstitial cells differentiate to activated myofibroblasts; results in stiffened and thickened aortic valve leaflets which impair efficient cardiac function.

7.1.2 The Aortic Valve

Located between the left ventricle of the heart and the ascending aorta, the aortic valve is composed of three semilunar valve cusps or leaflets (Fig. 7.1). In humans, each leaflet is usually less than 1 mm in thickness (Otto et al., 1994), while its composition is intricate and multi-layered. A single-cell layer of valvular endothelial cells (VECs) lines the leaflet aortic and ventricular surfaces. VECs are a distinctly different phenotype from endothelial cells that line the rest of the vasculature (Farivar et al., 2003; Butcher et al., 2004; Davies et al., 2004; Butcher et al., 2006; Young et al., 2007). VECs are oriented circumferentially on the valve leaflet, likely due to contact guidance from underlying collagen fibrils (Deck, 1986). This results in valve leaflet regions where VEC alignment and elongation are perpendicular to the dominant flow direction, in contrast to vascular endothelial cells that align parallel to flow. Of all active genes in both valvular and vascular endothelial cells in static culture, ~47% are unique to one of the two cell types (Farivar et al., 2003). VECs proliferate four times more rapidly than their vascular counterparts and prefer different to do so on different ECM substrates (Farivar et al., 2003). Shear stress-responsive genes (Butcher et al., 2006) and adhesion properties (Young et al., 2007) also differ between valvular and vascular ECs. Importantly, a large number of genes are differentially expressed between the aortic and ventricular sides of normal porcine aortic valve leaflets, implying the presence of phenotypic spatial heterogeneity within the valvular endothelial population itself (Simmons et al., 2005).

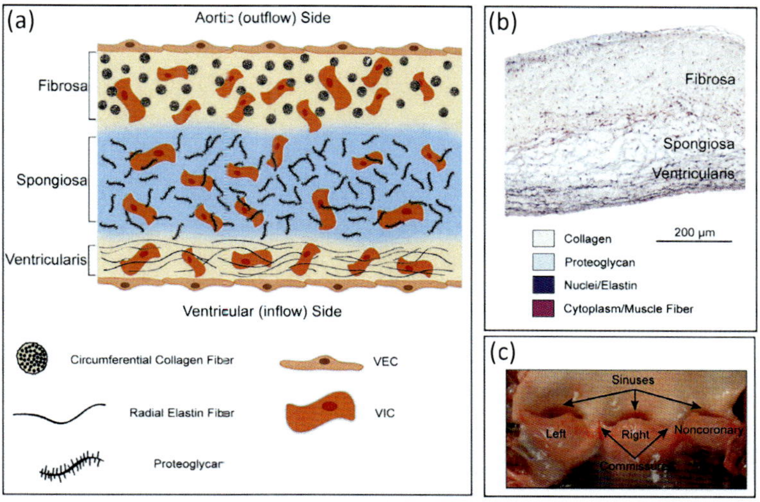

Figure 7.1 Aortic valve anatomy and morphology. (a) Schematic cross-sectional illustration of the tri-layer aortic valve with the aortic (outflow) surface on top. Valvular endothelial cells (VECs) coat the surfaces, while valvular interstitial cells (VICs) reside internally. The fibrosa is rich in circumferential collagen fibers, the spongiosa is primarily proteoglycan, and the ventricularis contains radial elastin. (b) Movat's Pentachrome histological stain of the porcine aortic valve cross section detailing layers and extracellular matrix composition. (c) Open view of the tri-leaflet porcine aortic valve displaying gross anatomy. Images (b) and (c) kindly provided by Krista L. Sider, University of Toronto.

The bulk of the valve consists of three distinct layers of cells and associated ECM components. The layer closest to the ventricular side of the leaflet is the ventricularis. This layer is rich in collagen and elastin, the latter of which is radially oriented and reduces radial strain during valve opening (Schoen, 2008). The middle "spongiosa layer" has an ECM with proteoglycans as the primary component. The spongiosa appears to act as a physical linking buffer between its neighboring layers as they undergo differential shear stresses and deformations during valve opening and closing (Sacks et al., 2009). Lastly, the fibrosa is located closest to the ascending aorta, and consists mainly of densely packed types I and III collagen (Thubrikar, 1990; Schoen, 1997). The fibrosa provides much of the valve's strength via circumferential collagen fiber orientation (Schoen, 2011). Valvular interstitial cells (VICs) are found throughout these three layers and mediate continuous homeostatic leaflet ECM remodeling through matrix component synthesis and matrix degrading enzyme expression (Taylor et al., 2003). They are a dynamic and heterogeneous cell population: healthy valves are composed primarily of quiescent fibroblasts, with a small portion (1–5%) of activated myofibroblasts (Pho et al., 2008). Porcine aortic valve leaflets also contain a large subpopulation of multipotent mesenchymal progenitor cells (capable of differentiating to the myofibrogenic, osteogenic, chondrogenic, and adipogenic lineages) (Chen et al., 2009). And while such a population has not been conclusively identified in the human valve, its existence is highly likely (Schoen, 2008).

The valve leaflets open and close with every heartbeat—more than three billion times over a human lifespan (Thubrikar, 1990). As the left ventricle contracts during systole, the aortic valve opens passively when left ventricular pressure is greater than that of the ascending aorta. This allows blood flow to the aorta and the systemic circulation. At the start of diastole, the left ventricle relaxes and fills with blood, causing the valve leaflets to co-apt and close when left ventricular pressure drops below that of the aorta (Thubrikar, 1990). A direct consequence of its position and function is that the aortic valve sits in a highly complex biomechanical environment. When open, the leaflet ventricular side experiences a substantial oscillatory hemodynamic shear stress (between –50 and 70 dynes/cm^2 (dPa) throughout the cardiac cycle (Yap et al., 2012b)), while the arterial side is presented with disturbed flow, recirculation zones, and eddy formation (Balachandran et al., 2011c), along with lower shear stresses of 1–3 dynes/cm^2 and a brief peak to 20 dynes/cm^2 during mid-systole (Yap et al., 2012a). While closed, the valve must withstand a substantial pressure load (80 mmHg in normotensive individuals, over 100 mmHg in type II hypertensive individuals (Butcher et al., 2008)) and the resultant tension to prevent retrograde flow. Simultaneously, the leaflets undergo large radial strains to co-apt and seal off the aorta. Lastly, they undergo high flexural stresses and deformations during the opening and closing motions (Merryman et al., 2006a; Merryman, 2007; Merryman et al., 2007; Butcher et al., 2008).

7.1.3 Aortic Valve Disease

In broad terms, the diseased aortic valve becomes stiff and thickened as it undergoes fibrosis (valvular sclerosis) (Otto et al., 1994; Stewart et al., 1997). Later in disease progression, sizeable focal calcific nodules are apparent and the leaflets become increasingly calcified, leading to impaired leaflet motion and closure, valvular stenosis, and finally obstructed blood flow and normal cardiac function (Fig. 7.2) (Otto et al., 1994; O'Brien et al., 1996; Otto et al., 1999). There is a distinct side-dependency to this pathology whereby the valve leaflet fibrosa (aortic side) is more prominently affected (Otto et al., 1994; O'Brien et al., 1996). Traditionally, valvular disease development was believed to be the result of age-related passive degeneration—a postulation that has been overturned in recent years. Instead, the current understanding is that aortic valve disease is an active, cell-mediated process in which the summation of biochemical factors and mechanical forces drive VIC activation and differentiation (Rajamannan et al., 2011). Indeed, a variety of environmental factors can induce activation of the quiescent VIC fibroblast fraction, leading to prototypical myofibroblast marker α-smooth muscle actin (α-SMA) expression, intracellular stress fiber development, and a resultant differentiation to myofibroblasts (Rabkin-Aikawa et al., 2004; Merryman et al., 2006c) that comprise approximately 30% of the diseased valve cellular population (Taylor et al., 2000). This myofibroblast differentiation pushes a healthy valve's ECM turnover towards pathological accumulation of poorly organized collagen and elastin, leading to leaflet thickening and stiffening (Otto et al., 1994; Fondard et al., 2005). This fibrotic replacement of healthy ECM is due in part to elevated collagen synthesis (Eriksen et al., 2006), but also a disruption of the homeostatic balance between the primary ECM remodeling enzymes (matrix metalloproteinases, MMPs), and their inhibitors (tissue inhibitors of metalloproteinases, TIMPs). While levels and activity of both enzyme families are upregulated in calcific aortic valve disease (CAVD), the balance becomes tilted towards MMPs with inadequate inhibition by TIMPs, resulting in both disorganized ECM fiber deposition and fragmentation of pre-existing elastin and collagen (Satta et al., 2003; Balachandran et al., 2009). Accompanying these occurrences are chronic inflammation (Otto et al., 1994), increased lipoprotein deposition (Warren and Yong, 1997), valve basement membrane disruption, and formation of nodules of ectopic bone and cartilage (Otto et al., 1994). In combination, these developments produce a severely altered biochemical and mechanical environment which has drastic effects on valvular function, namely impaired leaflet co-aptation and disturbed hemodynamic flow (Warren and Yong, 1997).

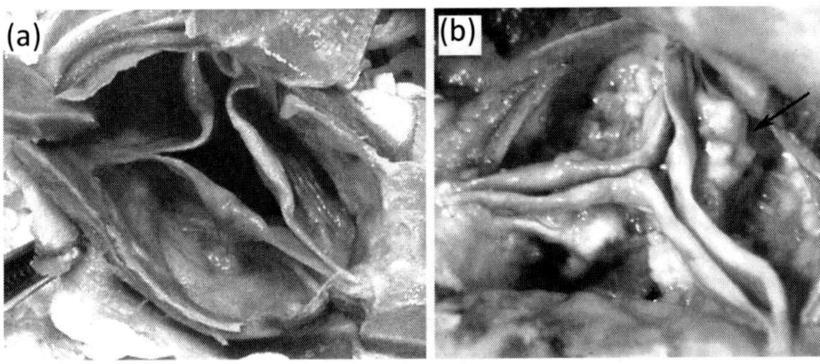

Figure 7.2 Gross aortic valve specimens, viewed from the ascending aorta above the heart. (a) Normal aortic valve leaflets appear thin, pliable, and open easily during systole. (b) A severely stenotic aortic valve displaying substantial lipid and calcific deposits (arrow) on the aortic side of the leaflets resulting in impaired cardiac function. (a) and (b), images with permission from Freeman RV, Otto CM. Spectrum of calcific aortic valve disease: pathogenesis, disease progression, and treatment strategies. *Circulation* 2005; **111**(24):3316–3326.

CAVD covers the spectrum of these valvular dysfunctions, from mild aortic valve sclerosis to severe aortic valve stenosis (Peltonen et al., 2007; Rajamannan et al., 2011). Indeed, 25% of the population over age 65 is affected by aortic valve sclerosis, while aortic valve stenosis affects 2–4% of those over age 84 (Otto et al., 1999). Aortic valve sclerosis is associated with a 50% increased risk for other cardiovascular events (Barasch et al., 2006) and is directly responsible, or a contributing factor to, over 15,000 deaths per year in the United States alone (Roger et al., 2011). Currently, there are no effective medical therapies. Although a variety of vascular diseases can be successfully treated with statins, clinical trials on valvular stenosis/sclerosis treatment with lipid-lowering statins have had, at best, mixed results (Shavelle et al., 2002; Rosenhek et al., 2004; Chan et al., 2007; Rossebo et al., 2008). Surgical intervention is presently the only treatment option, but is a last resort, as replacement valves suffer from serious side effects and variable device lifespan (Mendelson and Schoen, 2006; Bonow et al., 2008).

Thus, valvular fibrosis is an increasingly prevalent, costly, and deadly condition, the etiology of which arises from a complex mix of biochemical and mechanical factors at the tissue, cellular, and subcellular levels. In this sense, it is well suited as a model to further explore the implications of molecular and cellular biomechanics in disease.

7.2 The Myofibroblast

7.2.1 Role in Disease and Repair

A key event in the onset and development of valvular fibrosis is VIC activation and their subsequent differentiation into myofibroblasts expressing intracellular α-SMA stress fibers, which initiate unchecked ECM synthesis and leaflet thickening. The myofibroblast was first discovered in 1971 by Gabbiani et al., who identified modified fibroblasts in dermal wounds which they described as "intermediate forms between fibroblasts and smooth muscle cells" (Gabbiani et al., 1971). In the years since, a number of studies have implicated this cell type as a key player in a wide variety of reparative and disease processes through the establishment of tension, contraction, and ECM remodeling (reviewed in Hinz, 2010, and Tomasek et al., 2002). During physiological remodeling in healthy tissue, the myofibroblast exhibits a transient presence—upon completion of wound closure, these cells disappear via large-scale apoptosis (Desmouliere et al., 1995). Although currently uncertain, it is believed that myofibroblast release from mechanical stress or increased cell–cell cadherin junction formation after wound closure is the source of the apoptotic signal (Carlson et al., 2003; Hinz et al., 2004). In vitro, de-differentiation (as evidenced by a loss of α-SMA expression exclusive of cell death) appears to occur upon transforming growth factor-β1 (TGF-β1) antagonist delivery or subsequent seeding on compliant matrices (Hinz, 2007).

While myofibroblasts aid in dermal wound closure, unregulated myofibroblastic activity that is not terminated via apoptosis leads to fibrotic ECM overproduction along with tissue deformation and contracture. These occurrences provide a foundation for fibrosis development in all the major organs.

7.2.2 Identification and Cellular Characteristics

The precursor cells from which myofibroblasts originate are as varied as the tissue and organ types in which they play a role. Indeed, myofibroblasts seem to be recruited from nearly any suitable local cell type. In most tissues, they are believed to arise predominantly from local fibroblasts, but osteoblasts, smooth muscle cells, mesenchymal stem cells, epithelial cells, chondrocytes, and others have also been shown to contribute (reviewed in Hinz et al., 2007, and Barnes and Gorin, 2011). Despite their broad origins, the precursor cell differentiation to a myofibroblastic phenotype follows a similar sequence in which the cells develop a contractile stress fiber apparatus containing actin and non-muscle myosin IIa and IIb (Bond et al., 2011). Meanwhile, they maintain their actively synthetic state, expressing excessive disorganized ECM—particularly collagen and fibronectin (Gabbiani et al., 1971; Hinz et al., 2001). Myofibroblasts with this contractile skeleton but lacking α-SMA expression are termed "proto-myofibroblasts," while the presence of α-SMA-positive stress fibers denotes a "differentiated myofibroblast." In contrast to fibroblasts, myofibroblasts exhibit direct connections with their neighbors through intercellular gap junctions, implying the possibility that multicellular contractile units are formed (Gabbiani et al., 1978; Spanakis et al., 1998; Tomasek et al., 2002). While smooth muscle cells (SMCs) also share α-SMA-positive stress fibers and gap junctions, differentiated myofibroblasts typically lack expression of desmin, heavy chain smooth muscle myosin, and smoothelin (Schurch et al., 2007).

Myofibroblastic phenotype differentiation is also accompanied by alterations in the way cells attach to their substrates (Fig. 7.3). In vitro, fibroblasts lacking stress fibers form only small, immature focal complexes (cell–matrix adhesions) on the order of 2–6 μm (Geiger et al., 2001), which attach the surrounding ECM to the cytoskeleton via integrin receptor binding. Proto-myofibroblasts exhibit similar "classical" focal adhesions (FAs) at the upper range of focal complex length scales (Dugina et al., 2001). In stark contrast stand the supermature FAs found in differentiated myofibroblasts, which grow far larger (8–30 μm) and stronger (exert four times greater stress; ~12 nN/μm^2 (kPa)) than classical FAs (Dugina et al., 2001; Hinz et al., 2003). These multi-molecular supermature

FA complexes contain over 50 types of proteins and form the basis for substrate stiffness and stress perception by myofibroblasts (Hinz, 2007). Supermature FA development in differentiated myofibroblasts depends on a stiffened ECM and on contractile cytoskeleton formation through α-SMA incorporation (Hinz et al., 2003; Hinz, 2006). In addition, supermature FAs show high focal adhesion kinase (FAK) levels (Hinz, 2006), whose downstream signaling has been implicated in regulating α-SMA expression in response to mechanical force application to the integrin receptor portion of FAs (Chan et al., 2009). In this way, the differentiated myofibroblast exhibits an extensive network of cell–matrix connections—placing it in intimate contact with the biomechanics and biochemistry of its surrounding environment (Hinz, 2010).

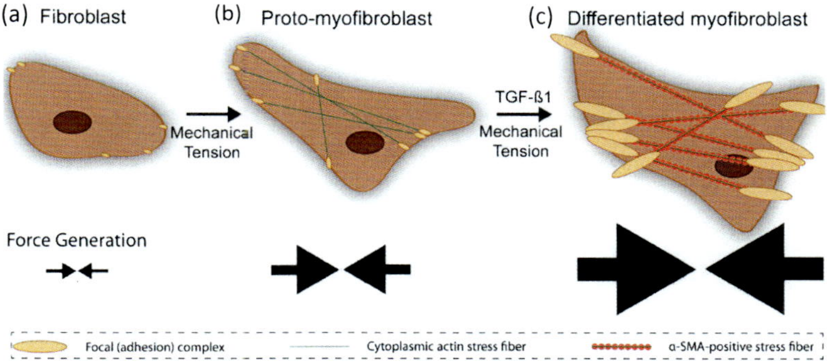

Figure 7.3 Progression of myofibroblast differentiation. (a) Fibroblasts or other local precursor cell types lack stress fibers and form only small, immature focal complexes (cell–matrix adhesions). (b) Mechanical tension induces the proto-myofibroblast to develop contractile stress fibers, allowing for increased force generation via larger, "classical" focal adhesions. (c) Further mechanical tension frees and activates TGF-β1, leading to the formation of a fully differentiated myofibroblast. Here, stress fibers contain α-SMA, supermature focal adhesions have formed, and fourfold increased stress is exerted by the cell on the surrounding matrix.

7.2.3 TGF-β1-Mediated Mechanical Control of Myofibroblast Differentiation

The primary cytokine responsible for inducing myofibroblast differentiation is believed to be the profibrotic growth factor

TGF-β1, whose downstream effects upon binding the TGFβ receptor type II (TGFβRII) are to upregulate α-SMA and collagen synthesis (Malmstrom et al., 2004). TGF-β1 is a key regulatory cytokine for a wide variety of processes both healthy and pathological. It plays a role in normal tissue repair and fibrosis development, mediates the inflammatory response and ECM production, and alters MMP and TIMP synthesis and activation (reviewed in (ten Dijke and Arthur, 2007)). Besides its effects on fibrosis, TGF-β1 also regulates proliferation of a wide variety of cell types, and as such general inhibition of TGF-β1 is recognized to be problematic for antifibrosis therapy due to the likelihood of widespread off-target effects (Varga and Pasche, 2008). TGF-β1 fibrotic effects are mediated primarily through the canonical Smad signaling pathway. In brief, active TGF-β1 binds to TGFβRII, which then recruits and phosphorylates the TGFβ receptor type I (TGFβRI). TGFβRII and TGFβRI form an activated complex containing a serine/threonine kinase, which phosphorylates the Smad2 and Smad3 proteins, which then bind Smad4 and translocate into the nucleus. There they act as transcription factors to induce a program of gene expression which promotes the fibrotic activities mentioned above (Heldin et al., 1997). Although there are some Smad-independent TGF-β1 effects, Smad activity is believed to be necessary for maximal myofibroblast differentiation and collagen synthesis (Schnabl et al., 2001).

In the context of myofibroblast differentiation, TGF-β1 induces α-SMA-positive stress fiber expression and thus a contractile phenotype through an elegant mechanosensitive pathway. In healthy tissue, resident cells are shielded from external stresses by a compliant, cross-linked ECM. Importantly, fibrotic scars and lesions exhibit substantial stiffness increases in comparison with the surrounding microenvironment (Tomasek et al., 2002). Myofibroblasts secrete a latent form of TGF-β1, which is inactivated while bound to two other proteins—LAP and latent TGF-β1-binding protein-1 (LTBP-1), which together form the large latent complex (Annes et al., 2004) (Fig. 7.4). LTBP-1 binds a variety of ECM proteins (fibrillin-1, fibronectin, vitronectin, etc.), creating a latent TGF-β1 pool in the ECM (Jenkins, 2008). Upon dissociation from LAP, TGF-β1 becomes biologically active; the mechanism of this dissociation is an inherently context-dependent phenomenon and

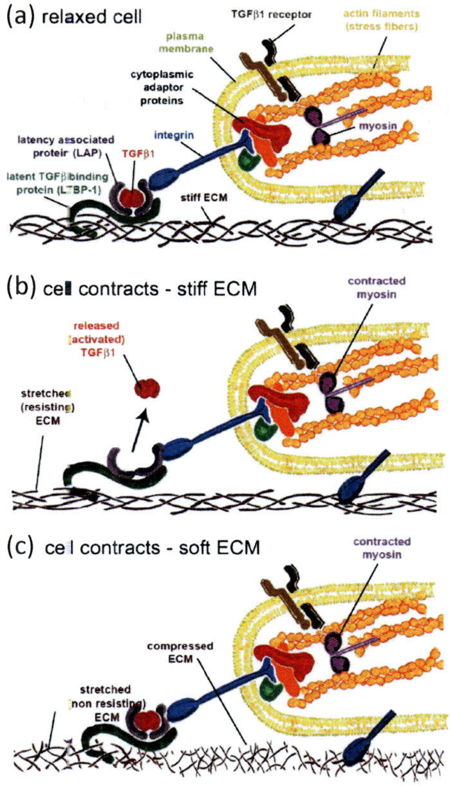

Figure 7.4 Integration of mechanics and biochemical signaling: integrin-mediated myofibroblast contraction activates TGF-β1. (a) Myofibroblasts secrete a latent TGF-β1 bound to LAP and LTPB-1 (together forming the large latent complex, LLC). Mechanical stress from myofibroblast contraction can be transmitted from the cytoskeleton to the LLC via integrins. (b) When the LLC is affixed to a stiff ECM, said ECM resists myofibroblast contractile forces. This "opens up" the LLC and releases biologically active TGF-β1, which then binds to the TGF-β1 receptor and acts through the canonical Smad signaling pathway to induce further α-SMA and collagen synthesis. (c) A compliant ECM does not provide sufficient mechanical resistance to induce conformational changes in LAP or LTBP-1 and release of TGF-β1; instead, the ECM is compressed and the LLC is pulled towards the myofibroblast. a, b and c, images reprinted from *Journal of Bodywork and Movement Therapies*, vol. 13, Wipff and Hinz, Myofibroblasts work best under stress, pp. 121–127, 2009, with permission from Elsevier.

can rely on proteolytic cleavage, interactions with thrombospondin-1 and/or mannose-6-phosphate receptor, and integrin activation (Jenkins, 2008). LAP and LTBP-1 contain the classical RGD integrin binding motif (Keski-Oja et al., 2004), and it appears that cell traction forces are transmitted through integrins directly to the large latent complex (Jenkins, 2008). Integrin activation of latent TGF-β1 independent of any proteolytic action was first demonstrated in the epithelial integrin $\alpha_v\beta_6$, which binds LAP but requires LTBP-1 for activation (Annes et al., 2004). Interestingly, purified integrin incubation with latent TGF-β1 is not enough to induce latent complex activation. In fact, a direct connection between integrins and the cytoskeleton is required. Treatment with a potent actin polymerization inhibitor (cytochalasin D) prevents activation, as does truncation of the cytoplasmic tail of integrin $\alpha_v\beta_6$ (Sheppard, 2005). Recently, Wipff et al. (2007) demonstrated a mechanism in myofibroblasts that is independent of proteolytic cleavage and demonstrates direct latent TGF-β1 activation by mechanical stress transmitted from the cytoskeleton through the myofibroblast integrins $\alpha_v\beta_5$ and $\alpha_v\beta_3$. They found that both external stretching and intracellular tension in myofibroblast cultures directly activate latent TGF-β1. Preventing LAP-integrin binding abolished TGF-β1 activation, while treatment with pharmacological agents to induce myofibroblast contraction (angiotensin-II, endothelin-1) increased TGF-β1 activation. In addition, α-SMA-positive stress fibers were required for this activation, as was a minimum threshold substrate stiffness of ≥5 kPa. At a stiffness of 47 kPa, basal TGF-β1 activation was increased 1.5-fold over the soft 5 kPa substrate. This implies that conformational change in the latent complex (thus releasing active TGF-β1) requires a minimum ECM stiffness/resistance against which the contractile myofibroblast can pull. It is of note that this 5 kPa threshold for latent TGF-β1 activation is below the threshold for α-SMA–positive stress fiber expression (~16 kPa; Goffin et al., 2006) and provides a putative means whereby mechanical stress TGF-β1 activation induces the progression of further tissue remodeling such that the ECM stiffens to the point of fibrotic α-SMA expression (Hinz, 2010).

7.2.4 Mechanical Control of Myofibroblast Differentiation Independent of TGF-β1

α-SMA expression and polymerization in stress fibers is also controlled by mixed mechanical/biochemical pathways besides that of TGF-β1. Of striking interest is the fact that TGF-β1 is only capable of inducing myofibroblast differentiation and upregulating α-SMA expression on stiff collagen substrates and not on compliant ones (Arora et al., 1999). Further studies have found that regardless of biochemical stimulation by TGF-β1, stiffness greater than 3 kPa (approximately that of healthy tissue) is required for proto-myofibroblast differentiation, where stress fibers lacking α-SMA form classical focal adhesions (Yeung et al., 2005a). For the development of differentiated myofibroblasts exhibiting α-SMA-positive stress fibers, substrate stiffness above 16 kPa (corresponding to that of contracting fibrotic or granulation tissue found in fibrotic scars) is necessary (Goffin et al., 2006). α-SMA mechanotransduction is regulated in a number of ways that terminate in control of α-SMA promoter activity. The small GTP-ase RhoA and its downstream effector Rho-associated protein kinase (ROCK) have been implicated as essential for stress-regulated α-SMA promoter activity (Zhao et al., 2007a). Many of the guanine nucleotide exchange factors (GEFs) and GTPase-activating proteins (GAPs) which function as molecular regulators of RhoA activity are intimately associated with the cytoskeleton. Thus their interaction with, and regulation of, RhoA are likely to be affected by cytoskeletal force transduction (Lessey et al., 2012). RhoA activation is required for long-term tension generation by myofibroblasts as it promotes actin-myosin contraction of the cytoskeleton through its myosin phosphatase inhibition and phosphorylated myosin light chain (MLC) maintenance (Katoh et al., 2001; Jenkins, 2008). In addition, force-induced activation of RhoA (and subsequently ROCK) at integrin-mediated force transfer sites (Zhao et al., 2007a) leads to phosphorylation of LIM kinase and cofilin, which increase actin filament assembly and filament stability (Bamburg et al., 1999). Importantly, the ratio between actin filaments and free actin monomers affects serum response factor (SRF) activity (Miralles et al., 2003). Wang et al. have shown that SRF binding to a CArG-B element in the α-SMA promoter region was a significant contributor to force-mediated α-SMA expression (Wang et al., 2002). More recently, Zhao et al. (2007a) found that force application via integrins increased α-SMA promoter activity by 3.5-fold and triggered nuclear translocation of the SRF transcriptional co-activator myocardin-related transcription factor-A (MRTF-A). Inhibition of MRTF-A nuclear translocation or of RhoA/ROCK abolished the force-mediated α-SMA increase. Interestingly, recent work found that TGF-β1-induced Smad3 activation can inhibit and delay the epithelial cell transition into myofibroblasts by MRTF-CArG interaction inhibition (Masszi et al., 2010) (further reviewed in (Masszi and Kapus, 2011)). Thus, a complete understanding of the role of Smad3 in myofibroblast differentiation regulation remains elusive.

Of late, Chan et al. proposed a mechanotransduction model which incorporates a number of these and other regulatory proteins into an overarching mechanism for control of actin assembly and α-SMA expression (Fig. 7.5) (Chan et al., 2010). They determined that force application via myofibroblast integrins induced FAK phosphorylation and activation and actin-binding protein gelsolin recruitment to the integrin. Force-induced α-SMA promoter activity was blocked by FAK knockdown or gelsolin deletion, while FAK knockdown also negatively regulated MRTF-A nuclear translocation. Gelsolin activation promoted actin stress fiber assembly, as did concomitant ROCK/LIM kinase/cofilin activity. In addition, they identified a role for another actin-binding protein, mammalian Diaphenous-related formin (mDia). mDia knockdown with siRNA partially abrogated force-mediated actin assembly, MRTF-A nuclear translocation, SRF expression, α-SMA promoter activation, and α-SMA expression (Chan et al., 2010). Together, these works demonstrate a system of co-existing and partially redundant proteins that regulate force-induced α-SMA expression.

Even expressed cytosolic α-SMA localization to the cytoskeleton proper is controlled by mechanical forces. When differentiated myofibroblasts are treated with ROCK inhibitors or transferred to soft culture substrates, α-SMA dislocation from the cytoskeletal β-actin filaments is observed (Goffin et al., 2006; Hinz, 2010). Although this phenomenon has not been explained mechanistically, it implies that (a) differentiated myofibroblast stress fibers are designed such that α-SMA removal does not cause them to be broken down, and (b) the contractile cytoskeleton must contain a mechanosensitive α-SMA "binding" element (Hinz, 2010). The evidence for such mechanosensitive binding is twofold. First, competitive delivery of an artificial peptide whose sequence matches that of the α-SMA N-terminus removes α-SMA from stress fibers (Hinz et al., 2002). Second, tension application has been shown to open "cryptic" cysteine residues on stress fibers which may act as putative α-SMA binding sites that are only accessible under cytoskeleton mechanical loading (Johnson et al., 2007).

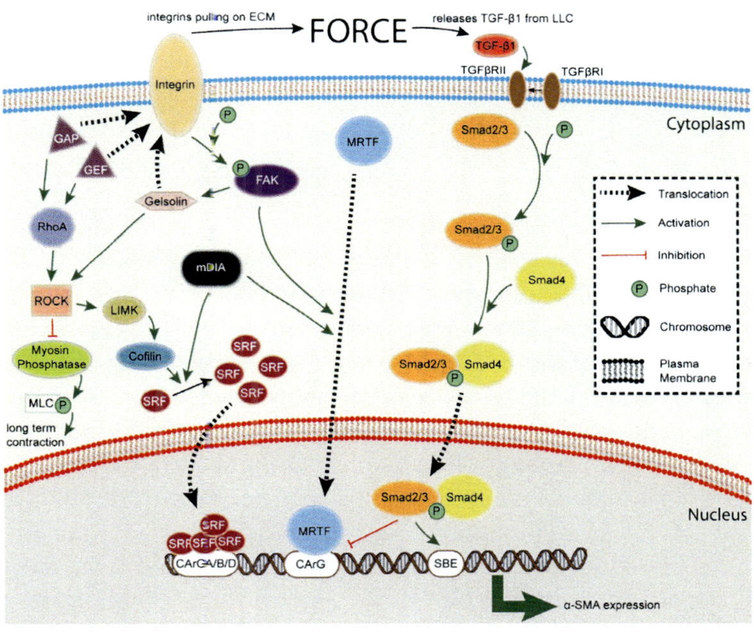

Figure 7.5 Myofibroblast mechanotransduction and α-SMA signaling. Mechanical tension and proto-myofibroblast contractile force release TGF-β1 from the large latent complex (LLC). TGF-β1 binds to the TGFβRI and II receptors and signals through the canonical Smad signaling pathway. Myofibroblast contraction results in force transfer via integrins, which leads to the activation of FAK, Gelsolin, and RhoA. Signaling from these molecules produces increased serum response factor (SRF) levels and induces SRF and MRTF translocation to the nucleus. There, together with molecules from the Smad pathway, they induce α-SMA expression.

7.3 Mechanical Regulation of Valvular Fibrosis

7.3.1 Substrate Stiffness and Cyclic Stretch

As is the case for myofibroblast differentiation in other body tissues, substrate stiffness and other mechanical stresses play a critical role in VIC function regulation. Plating porcine VICs on a rigid substrate produced myofibroblast differentiation as evidenced by α-SMA expression, while plating on collagen gels that were 10× softer prevented any myofibroblastic differentiation (Pho et al., 2008). Interestingly, when VICs were initially cultured and activated on high-modulus hydrogels, a decrease in substrate modulus via photodegradation resulted in myofibroblast de-activation. These results imply that valvular myofibroblast de-differentiation requires nothing more than a decrease in substrate elasticity (Kloxin et al., 2010). In addition, an exogenous tensile stress of 0.65 pN/μm^2 (Pa) applied directly to VICs with collagen-coated magnetite beads evoked ~60% higher α-SMA expression than in controls (Pho et al., 2008). Furthermore, RhoA effector protein cofilin expression increased threefold in myofibroblastic versus fibroblastic VIC populations. Cofilin knockdown slightly reduced α-SMA expression but almost completely abrogated cytoplasmic α-SMA localization to stress fibers and dramatically reduced myofibroblast-mediated collagen gel contraction. Additional ex vivo valve leaflet studies showed cofilin has low expression in healthy leaflets, but is strongly expressed and co-localized with α-SMA in diseased leaflets (Pho et al., 2008). Together, this work indicates that in VICs, cofilin marks myofibroblasts and is intimately connected with the cellular tension forces transmitted via stress fibers, which are required for VIC myofibroblast differentiation. Indeed, it offers evidence that the classical RhoA/ROCK pathway of α-SMA expression and myofibroblast differentiation is at work in the aortic valve. Interestingly, others have found that the HMG-CoA reductase inhibitor simvastatin dissipates calcific nodules formed by VICs on a variety of substrates by diminishing ROCK activity instead of the prototypical statin action mechanism (cholesterol biosynthesis prevention) (Monzack et al., 2009).

Studies by Yip et al. (2009) provided additional perspective on substrate stiffness-induced VIC calcification. Porcine aortic VICs grown in calcifying media on stiff collagen matrices developed calcified apoptotic cell aggregates, while aggregates on compliant matrices (approximately fourfold softer) were smaller though much more numerous. Compliant gel aggregates exhibited osteogenic calcification, with increased expression of the osteoblast transcripts Runx2, osteonectin, and osteocalcin. In contrast, stiff gel aggregates expressed high levels of stress fiber-localized α-SMA. Actin depolymerization by swinholide-A had no effect on compliant gel aggregate formation, but significantly reduced aggregate formation on stiff matrices. Consequently, it appears that myofibroblast-mediated cytoskeletal tension induces cellular contraction and aggregate formation on stiff gels. In addition, the Akt contraction-induced apoptosis pathway was activated, and a concomitant increase in apoptotic cells was observed upon release of constrained collagen gels seeded with VICs. See § 7.3.2 for further discussion.

Work by Merryman et al. (2006c) demonstrated that VIC stiffness is a function of cellular α-SMA content. They compared VICs isolated from the valves of the high-pressure left side of the heart (aortic and mitral; pressure gradients of 80 and 120 mmHg) versus those from the low-pressure right side (pulmonary and tricuspid; 10 and 25 mmHg). Micro-pipette aspiration of individual VICs found that left-side VICs were stiffer, while immunostaining showed these left-side VICs expressed more α-SMA and exhibited higher collagen biosynthesis levels. The same group demonstrated that basal VIC contraction in the aortic valve contributed a significant fraction of the aortic valve leaflet bending stiffness (Merryman et al., 2006b). As a whole, this work implies that the VIC response to local tissue stress increases is a concomitant rise in cellular stiffness mediated by α-SMA and collagen content increases, which equate to myofibroblast differentiation induction. Indeed, in aortic VICs induced to undergo said pathological differentiation through culture on stiff collagen matrices, increased cell stiffness correlated with rises in α-SMA expression (Wyss et al., 2012).

Unregulated ECM component synthesis is a clear hallmark of fibrotic valve disease. 20% stretch of VICs in vitro provokes significant collagen synthesis increases (as measured by [3H]-proline) and upregulation of the type III collagen gene COL3A1 (though not the type I collagen gene COL1A1) (Ku et al., 2006). In collagen gels seeded with VICs from the mitral valve, uni- or biaxial static tension increased glycosaminoglycan (GAGs synthesis and altered the proportion of distinctly sulfated GAGs (an indicator of compressive versus tensile loading stress) (Gupta et al., 2007). The importance of cyclic stretch on aortic valve disease and dysfunction has also been investigated by Balachandran and colleagues. An initial study utilized an ex vivo bioreactor to subject excised porcine aortic valve leaflet sections to cyclic circumferential stretch at a rate up to 300% s^{-1} and strain of 15% (Balachandran et al., 2006). The loading curve was near-physiological and closely mimicked the circumferential stress profile a valve leaflet would experience throughout the cardiac cycle in vivo. Peak wall shear stress in the bioreactor was estimated by computational fluid dynamics (CFD) to be only 3% of the in vivo case. After 48 h, stretched leaflets displayed increased collagen (~160%) and α-SMA (~55%) content when compared to static controls. In a follow-up study, the impact of cyclic stretch on aortic valve proteolytic enzymes was examined. The collagenase MMP-1 and gelatinases MMP-2/-9 were upregulated under cyclic stretch, as were cathepsin S and K (Balachandran et al., 2009). In addition, hyper-pathological stretch levels (20%) produced elevated cell proliferation and apoptosis. The MMP family of proteases are mediators of ECM degradation and remodeling, while cathepsins are elastolytic proteases, and all have been previously implicated in CAVD (Polyakova et al., 2004). Taken together, these results show that aortic valve cyclic stretching results in high levels of cell turnover and programs the ECM for degeneration and/or remodeling. Recent studies also identified upregulated bone morphogenic protein (BMP) expression in VICs from stenotic aortic valves (Yang et al., 2009). In further in vitro experiments, Balachandran et al. (2010) demonstrated that elevated cyclic stretch levels produced higher BMP-2/-4 and aortic valve calcification levels. Incubation with the BMP inhibitor noggin showed dose-dependent reductions in valvular calcium burden.

Other work has shown that VECs are responsive to a number of vasoactive signaling factors. Serotonin (5-HT) administration to whole leaflets induced a 25% decrease in cusp stiffness which was abrogated when 5-HT was co-administered with the endothelial nitric oxide synthase (eNOS) inhibitor L-NAME or mechanical endothelial denudation (El-Hamamsy et al., 2009). Cyclic stretch upregulates the 5-HT receptors 5-HT_{2A} and 5-HT_{2B} in VECs (Balachandran et al., 2011a), while 5-HT_{2A} inhibition abrogated 5-HT-mediated increases in collagen synthesis, cell proliferation, and cusp stiffness (Balachandran et al., 2011b). Meanwhile, endothelin-1 (ET-1) treatment increased leaflet stiffness by over a third, an effect that was absent when leaflets were treated with cytochalasin B, another actin-depolymerizing agent (El-Hamamsy et al., 2009). This effect is of significant importance, as ET-1 is an endothelium derived peptide expressed by VECs, and as such implies that ET-1 produced from VECs induces VIC stiffening via the VIC cytoskeleton. Together, these factors produce endothelium-dependent changes in aortic valve mechanical properties, which in turn alter the substrate stiffness seen by VICs, thereby altering VIC fate.

7.3.2 The Role of TGF-β1 in Aortic Valve Disease

Earlier the classical mechanisms underlying the central role of TGF-β1 in promoting myofibrogenesis were described. The importance of this molecule to fibrosis progression in the valvular microenvironment, with its multifaceted mechanical/biochemical/cellular/ECM composition, has therefore been the focus of substantial study. Here, key valve-specific experimental findings and their resultant implication of TGF-β1 as a potent mediator of VIC myofibroblastic differentiation and valvular pathogenesis are discussed. A decade ago, a study of human aortic valve cusps identified that there are significantly higher TGF-β1 levels in calcified aortic valve cusps than in non-calcified cusps (Jian et al., 2003). Calcified cusps exhibited consistently high latency associated peptide (LAP) levels, further implicating force-induced myofibrogenesis in CAVD etiology (Jian et al., 2003). Calcified cusps have also been shown to exhibit increased levels of alkaline phosphatase (a potent calcification inducer), MMP-9, and TIMPs (Clark-Greuel et al., 2007). TGF-β1 addition to VIC cultures induced aggregate formation, apoptosis, and calcification—concurrent treatment with an actin-depolymerizing agent prevented aggregate formation, while anti-apoptosis agents inhibited apoptosis and calcification (Jian et al., 2003).

Walker et al. (2004) demonstrated that TGF-β1 administration to cultured VICs invoked a dose-dependent α-SMA expression and stress fiber formation increase. Using a collagen matrix contraction assay, they also found that TGF-β1 and mechanical force induced actin fiber-dependent synergistic increases in VIC contractility. Subsequent to TGF-β1treatment, VIC myofibroblasts induced significant ECM fibronectin reorganization. Further studies indicated that the ECM composition can mediate TGF-β1-induced α-SMA expression (Cushing et al., 2005). Seeding VICs on fibronectin or heparin (which interact and bind with TGF-β1) produced higher α-SMA levels than controls, while a collagen-coated substrate exerted a protective effect. This collagen-based protection has also been demonstrated on VIC calcification, but the underlying mechanism remains to be elucidated (Rodriguez and Masters, 2009). Work by others has shown that in VICs, TGF-β1 induces myofibroblast differentiation in a matrix stiffness-dependent manner (Pho et al., 2008). Importantly, both endogenous force generated by cellular contraction against a stiff ECM and exogenous force applied with collagen-coated magnetite beads induced α-SMA expression increases. Others have shown that TGF-β1 induces MMP-9 and alkaline phosphatase expression along with increased apoptosis (Clark-Greuel et al., 2007).

Along with their conclusions that ECM stiffness affects calcific aggregate formation in VICs, Yip et al. (2009) explored the effect of ECM stiffness on VIC myofibrogenesis. TGF-β1 addition to cultures resulted in increased aggregate formation on stiff matrices through α-SMA expression induction and enhanced cellular contractility. Compliant matrices were insensitive to TGF-β1-induced aggregate formation as well as aggregate reduction by actin depolymerization. Most interestingly, while endogenous TGF-β1 levels did not differ between substrates, the TGFβ receptor type I (TGFβRI) was expressed at significantly lower levels on compliant matrices than stiff ones, suggesting VIC responsiveness to myofibroblast differentiation may be mediated by matrix stiffness-dependent TGFβRI expression.

The Smad signaling pathway has also been implicated in VIC differentiation to activated myofibroblasts. TGF-β1 induces Smad activation and nuclear Smad-3 localization in VICs, while basic fibroblast growth factor (FGF-2) inhibits TGF-β1 induced myofibroblast differentiation by blocking Smad protein nuclear translocation (Cushing et al., 2008). VIC cultures treatment with FGF-2 reduced α-SMA expression and blocked TGF-β1-mediated myofibrogenesis. In addition, FGF-2 inhibited TGF-β1-mediated cellular contraction and aggregate formation. Others have shown that TGF-β1 regulates VIC activation and monolayer wound repair through Smad2/3 phosphorylation and resultant α-SMA expression increase (Liu and Gotlieb, 2008). In VICs, the canonical Wnt pathway crosstalks with TGF-β/Smad signaling, as TGFβRI kinase activation induces nuclear β-catenin translocation and myofibroblast differentiation (Chen et al., 2011). Interestingly, this occurs only on matrices with stiffness similar to that of the fibrosa (disease-prone) valve layer and not on softer matrices with a ventricularis-like stiffness. β-catenin degradation prevented myofibroblast differentiation without affecting Smad2/3 activity. While β-catenin and Wnt3A treatment alone did not induce myofibrogenesis, TGF-β1 and Wnt3A together produced synergistic increases in myofibroblast differentiation. In diseased valve fibrosa, myofibroblasts colocalized with increased Wnt3A, β-catenin, TGF-β1, and activated Smad2/3 (Chen et al., 2011). Stiffness-dependent crosstalk between the TGF-β1 and Wnt signaling pathways may therefore make the stiffer fibrosa layer more prone to disease.

Inhibition of TGF-β1-mediated myofibroblast differentiation has also been achieved by treating VICs with C-type natriuretic peptide (CNP). CNP-treated VICs exhibited reduced α-SMA expression and myofibroblast-mediated gel contraction, which was abrogated by siRNA knockdown of the CNP receptor NPR-B (Yip et al., 2011). Consistent with this α-SMA expression inhibition, cultured VIC treatment with CNP prevented cellular stiffness increases that were otherwise induced by stiff substrates (Wyss et al., 2012). While specific mechanisms of action remain to be elucidated, other natriuretic peptide family members have been shown to inhibit fibrosis via protein kinase G (PKG) activation and phosphorylated Smad3 inhibition (Li et al., 2008b). Interestingly, CNP is more highly expressed on the disease-protected ventricular side of the aortic valve leaflets than on the disease-prone aortic side (Simmons et al., 2005). In addition, expression of CNP, its activating enzyme furin, and NPR-B are all dramatically reduced in diseased aortic valves (Peltonen et al., 2007). Consequently, CNP appears to play an important role in regulating valve homeostasis and maintaining a quiescent VIC population, perhaps by counteracting TGF-β1-induced Smad signaling.

While cyclic stretch alone has been shown to be a potent inducer of VIC myofibroblast differentiation, it has also been shown to act synergistically with TGF-β1. These synergistic effects were observed at 15% strain, which induced significant increases in α-SMA expression and collagen biosynthesis (Merryman et al.,

2007). This implies that cyclic stretch may free latent TGF-β1, producing a bioactive form which then activates VICs and leads to myofibroblast differentiation (Merryman, 2008). Stretch-activation of TGF-β1 may be particularly potent in the disease-prone fibrosa layer of aortic valve leaflets, where gene expression studies suggest there is an abundance of TGF-β1 available for bioactivation (Simmons et al., 2005; Merryman, 2010).

7.4 Conclusions

Fibrosis onset is characterized by deregulation of ECM homeostasis in which excess disorganized, fibrous collagen is secreted and produces stiffened scar tissue. In the aortic valve, this same deregulation leads to the activation of a quiescent fibroblast population, which subsequently differentiates into myofibroblasts exhibiting α-SMA-containing, contractile stress fibers. This differentiation is highly dependent upon valve stiffness and is mediated by a variety of mechanical forces including cyclic stretch and fluid shear stress. Biochemical signaling is also highly involved in this disease process, and a complex crosstalk between chemical and mechanical factors exists (Table 7.2). Aortic valve fibrosis development severely hinders its normal function and has far-reaching effects on the cardiovascular system.

Currently, therapeutic options for those who suffer from advanced aortic valve disease are limited. Valve replacement therapy has consistently improved over the past few decades, but current options are still lacking—mechanical valves require the recipient to receive lifelong anticoagulant therapy, while bioprosthetic valves (typically manufactured from porcine aortic valves or bovine pericardial tissue) have relatively limited durability and often require replacement surgeries (Bonow et al., 2008). Pharmaceutical intervention has been, to date, largely ineffective. A number of the drug treatments that have been found to combat other vascular diseases have been tested for valve disease treatment, but to no avail. In the past few years, in vitro work and retrospective clinical trials have implicated statins (popular and effective cholesterol-lowering drugs which are effective at treating atherosclerosis) as putative treatments (Shavelle et al., 2002; Rosenhek et al., 2004; Monzack et al., 2009). Unfortunately, prospective clinical trials found no statistically significant benefit for treating aortic valve disease (Chan et al., 2007; Rossebo et al., 2008). In light of the clear effect of mechanical forces on the aortic valve, future therapeutic efforts will need to account for, and indeed embrace, the biomechanical and biochemical stimuli that together are responsible for aortic valve disease pathogenesis. Putative therapeutics may target disruption of myofibroblast contraction mechanisms and/or aim to prevent ECM stiffening in an effort to short-circuit this pathological positive feedback loop. Such efforts may be achieved by inhibition of the Smad, RhoA/ROCK, or integrin-dependent signaling pathways.

Table 7.2 Key molecular players in valvular fibrosis

Molecule	Description	Effect*
5-HT	5-hydroxytryptamine (serotonin); reduces leaflet stiffness	+
5-HT$_{2A/2B}$	Serotonin receptors; G-protein coupled; upregulated by cyclic stretch	+
α-SMA	α-smooth muscle actin; prototypical myofibroblast marker; key component of contractile stress fibers	–
β-catenin	Induces α-SMA; nuclear translocation activated by Wnt3A and TGF-β1; upregulated in fibrotic valves	–
ALP	Alkaline phosphatase; inducer of osteoblastic calcification; upregulated by TGF-β1 and in calcified leaflets	–
BMP-2/-4	Bone morphogenic proteins; induce leaflet calcification; upregulated by cyclic stretch	–
CArG	Binding element in α-SMA promoter	–
Cathepsin S/K	Elastolytic proteases; upregulated by cyclic stretch	–
CNP	C-type natriuretic peptide; inhibits fibrosis via PKG; downregulated in fibrotic valves	+
Cofilin	Activated by LIM kinase; assembles and stabilizes actin filaments; upregulated in fibrotic leaflets	–
Collagen	Key component of ventricularis and fibrosa; strong in tension; over-expressed and disorganized in fibrosis	+/–

(*Continued*)

Table 7.2 *(Continued)*

Molecule	Description	Effect*
Cytochalasin B/D	Actin-depolymerizing agents	+
Elastin	Key component of ventricularis; reduces radial valvular strain; over-expressed and disorganized in fibrosis	+/-
eNOS	Endothelial nitric oxide synthase; produces nitric oxide (NO); downstream effector of 5-HT	+
ET-1	Endothelin-1; produced by VECs; upregulates TGF-β1; increases leaflet stiffness	-
FAK	Focal adhesion kinase; upregulated in supermature FAs; mediates force-induced α-SMA-promoter activation	-
FGF-2	Basic fibroblast growth factor-2; inhibits TGF-β1-mediated myofibrogenesis and blocks Smad translocation	+
GAP	GTPase-activating protein; inhibitory regulator of RhoA; bound to cytoskeleton	+
GEF	Guanine nucleotide exchange factor; excitatory regulator of RhoA; bound to cytoskeleton	-
Gelsolin	Actin binding protein; required for force-induce α-SMA expression; recruited to integrin by force application	-
Integrins	Transmembrane receptors; cell–cell or cell-ECM; key transducers of mechanical force	-
LAP	Latency associated peptide; component of LLC, upregulated in calcified leaflets	-
LIM kinase	Activated by ROCK, phosphorylates cofilin; results in actin assembly and stability	-
LLC	Large latent complex; composed of LAP, LTBP-1, and TGF-β1; inactivated integrin-bound TGF-β1	-
L-NAME	N-nitro-L-arginine-methyl-ester; inhibitor of eNOS	-
LTBP-1	Latent TGF-β1 binding protein-1; component of LLC; RGD domain binds LLC to integrins	-
mDia	Mammalian diaphanous-related formin; mediates force-induced MRTF-A/SRF/α-SMA promoter activity	-

Molecule	Description	Effect*
MLC	Myosin light chain; activation via RhoA promotes long-term generation of cellular tension	–
MMP-1/-2/-9	Matrix metalloproteinases; degrade and remodel ECM; upregulated by TGF-β1 and cyclic stretch	–
MRTF-A	Myocardin-related transcription factor-A; SRF transcriptional co-activator	–
NPR-B	Natriuretic peptide receptor-B; CNP receptor; downregulated in fibrotic valves	+
PKG	Protein kinase-G; inhibits nuclear translocation of phosphorylated Smad3	+
RhoA	Small GTP-ase; key effector of α-SMA mechanotransduction by MLC and ROCK	–
ROCK	Rho-associated protein kinase; downstream effector of RhoA; upregulates α-SMA	–
Wnt3A	Induces β–catenin nuclear translocation; upregulated in fibrotic valves	–
(Simva)statin	HMG-CoA reductase inhibitor, reduces ROCK activity to dissipate calcific nodules	+
Smad2/3/4	Downstream effectors of canonical TGF-β1 signaling; upregulated in fibrotic leaflets	–
SRF	Serum response factor; binding to CArG contributes to force-mediated α-SMA expression	–
Swinholide-A	Actin-depolymerizing agent	+
TGF-β1	Transforming growth factor-β1; induces α-SMA expression by TGFβRI/II and Smads	–
TGFβRI/II	TGFβ receptors; serine/threonine kinase; signal via Smads to upregulate α-SMA	–
TIMPs	Tissue inhibitors of metalloproteinases; inhibit MMPs; upregulated by TGF-β1	+

*+, protective; –, pathogenic.

While we have come a long way in understanding the mechanics-mediated pathobiology of valvular fibrosis, there is much that remains a mystery. While the mechanical properties of normal aortic valve tissue have been widely reported, biomechanical characterization of the valvular fibrosis produced by animal valve disease models is lacking (Sider et al., 2011) and may be aided by the development of novel tools and techniques for in vivo or ex vivo tissue mechanical property measurement (Nadkarni et al., 2005; Krishnamurthy et al., 2011; Zhao et al., 2011; Sewell-Loftin et al., 2012). In addition, the relative contributions of fibrosis and calcification to leaflet mechanics alterations in vivo are unknown, as are the relative magnitudes of the resultant feed-forward leaflet stiffening effects on further fibrotic or calcific progression. Even in early-stage valve disease there exists both prominent mineralization and myofibroblastic differentiation (Otto et al., 1994), and thus to date it has been difficult to separately study how these concurrently developing pathogenic mechanisms contribute to disease initiation and progression. Of particular importance is that the earliest initiation events in CAVD are drastically understudied. Improved understanding of the way in which a wide variety of early pathological insults to the valve leaflets combine and lead to later-stage myofibroblast differentiation is sorely needed and may reveal opportunities for diagnosis and therapeutic intervention prior to the onset of large-scale myofibroblast differentiation and unchecked fibrosis. While novel, well-characterized animal models will be essential to these investigations, the development of in vitro culture systems that better mimic the in vivo valve state also will be useful to accomplishing these goals. Instead of examining single types of mechanical force and solitary biochemical stimuli, future studies into aortic valve disease will require integration of multiple mechanical force types (substrate stiffness, fluid shear stress, cyclic stretch, etc.) and a variety of biochemical signaling pathways (TGF-β1/Smad, RhoA/ROCK, etc., and all their downstream effectors) in order to better mimic the valvular microenvironment complexity (Butcher et al., 2008). In vitro modeling of the complex valvular microenvironment will be facilitated by emerging "organ-on-a-chip" platforms (Huh et al., 2012) that reconstitute organ level functions by integrating many of the cell types, ECM components, and in some cases, mechanical forces (Moraes et al., 2010; Srigunapalan et al., 2011) that comprise the valvular microenvironment. These microtechnologies offer the potential for high-throughput, combinatorial investigations akin to those in genomics and proteomics, and combining them with systems biology and modeling approaches may provide insights into the complex mechanobiological interactions and regulatory pathways that drive the earliest stages of valve pathogenesis. Such investigations are key to furthering our understanding of how this complex and dynamic system responds to a wide variety of stimuli.

Problems

7.1 Table 7.1 briefly describes the fibrotic/foreign body response to implanted biomaterials/biomedical implants. Many applications including orthopedics, dental implants, pacemakers, biosensors, and drug delivery devices require the use of stiff materials of construction, which can exacerbate the fibrotic response.

(a) Propose and defend a method to mitigate fibrous encapsulation around a stiff biomedical implant.

(b) Fibrous encapsulation is typically an undesired medical development. For example, encapsulation of glucose sensors used in the management of diabetes can greatly reduce glucose transport to the sensor, resulting in inaccurate readings. Identify and justify one medical application where fibrous encapsulation via a foreign body response would be a desired and beneficial result to ensure proper implant function.

7.2 Identify three molecules/pathways that may be poor targets for pharmaceutical intervention to combat valvular fibrosis, due to the possibility for broad off-target effects throughout the rest of the body.

7.3 A bicuspid aortic valve (BAV) is a congenital malformation of the aortic valve where two valve leaflets are fused together during embryonic development. The valve therefore has only two leaflets, instead of the normal three. BAVs are more prone to early development of fibrosis, stiffening, and calcification than normal tricuspid (three-leaflet) aortic valves (TAV). Propose one reason why BAVs are more prone to early fibrosis than TAVs.

7.4 To determine the importance of the large latent complex to biomechanical signaling in VICs, a genetic mutation is introduced such that the RGD sequences in LAP and LTBP-1 are disrupted. These genetically modified VICs are seeded on collagen-coated petri dishes and are treated with angiotensin-II, a pharmacological agent that induces cellular contraction. Describe what result you expect this mutation of RGD to have on the synthesis of α-SMA by these VICs and your reasoning therein.

7.5 In your own words, describe the downstream effects and end result of a pharmacological inhibition of RhoA in cultured VICs.

7.6 Choose one of the molecules in Table 7.2. Using additional research sources, write a short essay (300–500 words) describing the molecule, including information on its structure, function, and regulatory signaling mechanisms.

References

Aarssen, L. W., B. S. Schamp and J. Pither (2006). Why Are There So Many Small Plants? Implications for Species Coexistence. *J. Ecol.* **94** (3): 569–580.

Abe, I., N. Ochiai, H. Ichimura, A. Tsujino, J. Sun and Y. Hara (2004). Internodes Can Nearly Double in Length with Gradual Elongation of the Adult Rat Sciatic Nerve. *J. Orthop. Res.* **22** (3): 571–577.

Abe, I., A. Tsujino, Y. Hara, H. Ichimura and N. Ochiai (2002). Paranodal Demyelination by Gradual Nerve Stretch Can Be Repaired by Elongation of Internodes. *Acta Neuropathol. (Berl)* **104** (5): 505–512.

Abe, I., A. Tsujino, Y. Hara and N. Ochiai (2003). Effect of the Rate of Prestretching a Peripheral Nerve on Regeneration Potential after Transection and Repair. *J. Orthop. Sci.* **8** (5): 693–699.

Abrams, R. A., J. M. Butler, S. Bodine-Fowler and M. J. Botte (1998). Tensile Properties of the Neurorrhaphy Site in the Rat Sciatic Nerve. *J. Hand Surg. Am.* **23** (3): 465–470.

Ackerley, S., P. Thornhill, A. J. Grierson, J. Brownlees, B. H. Anderton, P. N. Leigh, et al. (2003). Neurofilament Heavy Chain Side Arm Phosphorylation Regulates Axonal Transport of Neurofilaments. *J. Cell Biol.* **161** (3): 489–495.

Adrian, M., J. Dubochet, J. Lepault and A. W. McDowall (1984). Cryo-Electron Microscopy of Viruses. *Nature* **308** (5954): 32–36.

Aebi, U., J. Cohn, L. Buhle and L. Gerace (1986). The Nuclear Lamina Is a Meshwork of Intermediate-Type Filaments. *Nature* **323** (6088): 560–564.

Ahern, K. (2013) July 31, 2013. http://www.youtube.com/watch?v=teV62zrm2P0.

Ahmad, F. J., C. J. Echeverri, R. B. Vallee and P. W. Baas (1998). Cytoplasmic Dynein and Dynactin Are Required for the Transport of Microtubules into the Axon. *J. Cell Biol.* **140** (2): 391–401.

Ahmad, F. J., J. Hughey, T. Wittmann, A. Hyman, M. Greaser and P. W. Baas (2000). Motor Proteins Regulate Force Interactions between Microtubules and Microfilaments in the Axon. *Nat. Cell Biol.* **2** (5): 276–280.

Akey, C. W. (1989). Interactions and Structure of the Nuclear Pore Complex Revealed by Cryo-Electron Microscopy. *J. Cell Biol.* **109** (3): 955–970.

Akinc, A., A. Zumbuehl, M. Goldberg, E. S. Leshchiner, V. Busini, N. Hossain, et al. (2008). A Combinatorial Library of Lipid-Like Materials for Delivery of RNAi Therapeutics. *Nat. Biotechnol.* **26** (5): 561–569.

Allen, K. B., F. M. Sasoglu and B. E. Layton (2009a). Cytoskeleton-Membrane Interactions in Neuronal Growth Cones: A Finite Analysis Study. *J. Biomech. Eng.* **131** (2): 021006.

Allen, K. B., F. M. Sasoglu and B. E. Layton (2009b). Cytoskeleton-Membrane Interactions in Neuronal Growth Cones: A Finite Analysis Study. *J. Biomechan. Eng.-Trans. ASME* **131** (2): 1–10.

Almqvist, N., R. Bhatia, G. Primbs, N. Desai, S. Banerjee and R. Lal (2004). Elasticity and Adhesion Force Mapping Reveals Real-Time Clustering of Growth Factor Receptors and Associated Changes in Local Cellular Rheological Properties. *Biophys. J.* **86** (3): 1753–1762.

Andersen, S. S. and G. Q. Bi (2000). Axon Formation: A Molecular Model for the Generation of Neuronal Polarity. *Bioessays* **22** (2): 172–179.

Anderson, D. G. (2008). A Combinatorial Library of Lipid-Like Material for Delivery of RNAi Therapeutics. *Technol. Rev.* **111** (4): 93.

Anderson, J. D. and J. Widom (2001). Poly(Da-Dt) Promoter Elements Increase the Equilibrium Accessibility of Nucleosomal DNA Target Sites. *Mol. Cell. Biol.* **21** (11): 3830–3839.

Andreassen, T. T., K. Seyer-Hansen and H. Oxlund (1981). Biomechanical Changes in Connective Tissues Induced by Experimental Diabetes. *Acta Endocrinol. (Copenh)* **98** (3): 432–436.

Annes, J. P., Y. Chen, J. S. Munger and D. B. Rifkin (2004). Integrin Alphavbeta6-Mediated Activation of Latent TGF-Beta Requires the Latent TGF-Beta Binding Protein-1. *J. Cell Biol.* **165** (5): 723–734.

Applewhite, D. A., K. D. Grode, D. Keller, A. Zadeh, K. C. Slep and S. L. Rogers (2010). The Spectraplakin Short Stop Is an Actin-Microtubule Cross-Linker That Contributes to Organization of the Microtubule Network. *Mol. Biol. Cell* **21** (10): 1714–1724.

Arai, N., K. Yasuoka and Y. Masubuchi (2007). Spontaneous Self-Assembly Process for Threadlike Micelles. *J. Chem. Phys.* **126** (24): 244905.

Araya, C. L. and D. M. Fowler (2011). Deep Mutational Scanning: Assessing Protein Function on a Massive Scale. *Trends Biotechnol.* **29** (9): 435–442.

Archer, D. R., L. B. Dahlin and W. G. McLean (1994). Changes in Slow Axonal Transport of Tubulin Induced by Local Application of Colchicine to Rabbit Vagus Nerve. *Acta Physiol. Scand.* **150** (1): 57–65.

Arora, P. D., N. Narani and C. A. McCulloch (1999). The Compliance of Collagen Gels Regulates Transforming Growth Factor-Beta Induction of Alpha-Smooth Muscle Actin in Fibroblasts. *Am. J. Pathol.* **154** (3): 871–882.

Ashkin, A. and J. M. Dziedzic (1987). Optical Trapping and Manipulation of Viruses and Bacteria. *Science* **235** (4795): 1517–1520.

Atilgan, E., D. Wirtz and S. X. Sun (2006). Mechanics and Dynamics of Actin-Driven Thin Membrane Protrusions. *Biophys. J.* **90** (1): 65–76.

Baas, P. W. (2000). Strategies for Studying Microtubule Transport in the Neuron. *Microsc. Res. Tech.* **48** (2): 75–84.

Baas, P. W. and M. M. Black (1990). Individual Microtubules in the Axon Consist of Domains That Differ in Both Composition and Stability. *J. Cell Biol.* **111** (2): 495–509.

Baas, P. W., C. Vidya Nadar and K. A. Myers (2006). Axonal Transport of Microtubules: The Long and Short of It. *Traffic* **7** (5): 490–498.

Badoual, M., F. Julicher and J. Prost (2002). Bidirectional Cooperative Motion of Molecular Motors. *Proc. Natl. Acad. Sci. U. S. A.* **99** (10): 6696–6701.

Bajaj, P., X. Tang, T. A. Saif and R. Bashir (2010). Stiffness of the Substrate Influences the Phenotype of Embryonic Chicken Cardiac Myocytes. *J. Biomed. Mater. Res. Part A* **95A** (4): 1261–1269.

Balachandran, K., M. A. Bakay, J. M. Connolly, X. Zhang, A. P. Yoganathan and R. J. Levy (2011a). Aortic Valve Cyclic Stretch Causes Increased Remodeling Activity and Enhanced Serotonin Receptor Responsiveness. *Ann. Thorac. Surg.* **92** (1): 147–153.

Balachandran, K., S. Hussain, C. H. Yap, M. Padala, A. H. Chester and A. P. Yoganathan (2011b). Elevated Cyclic Stretch and Serotonin Result in Altered Aortic Valve Remodeling Via a Mechanosensitive 5-Ht(2a) Receptor-Dependent Pathway. *Cardiovasc Pathol* **21** (3): 206–213.

Balachandran, K., S. Konduri, P. Sucosky, H. Jo and A. P. Yoganathan (2006). An ex vivo Study of the Biological Properties of Porcine Aortic Valves in Response to Circumferential Cyclic Stretch. *Ann. Biomed. Eng.* **34** (11): 1655–1665.

Balachandran, K., P. Sucosky, H. Jo and A. P. Yoganathan (2009). Elevated Cyclic Stretch Alters Matrix Remodeling in Aortic Valve Cusps:

Implications for Degenerative Aortic Valve Disease. *Am. J. Physiol. Heart Circ. Physiol.* **296** (3): H756–H764.

Balachandran, K., P. Sucosky, H. Jo and A. P. Yoganathan (2010). Elevated Cyclic Stretch Induces Aortic Valve Calcification in a Bone Morphogenic Protein-Dependent Manner. *Am. J. Pathol.* **177** (1): 49–57.

Balachandran, K., P. Sucosky and A. P. Yoganathan (2011c). Hemodynamics and Mechanobiology of Aortic Valve Inflammation and Calcification. *Int. J. Inflam.* **2011**: 263870.

Bamburg, J. R. (1999). Proteins of the Adf/Cofilin Family: Essential Regulators of Actin Dynamics. *Annu. Rev. Cell Dev. Biol.* **15**: 185–230.

Bamburg, J. R., A. McGough and S. Ono (1999). Putting a New Twist on Actin: Adf/Cofilins Modulate Actin Dynamics. *Trends Cell Biol.* **9** (9): 364–370.

Bandak, F. A. (1995). On the Mechanics of Impact Neurotrauma: A Review and Critical Synthesis. *J. Neurotrauma* **12** (4): 635–649.

Barasch, E., J. S. Gottdiener, E. K. Marino Larsen, P. H. Chaves and A. B. Newman (2006). Cardiovascular Morbidity and Mortality in Community-Dwelling Elderly Individuals with Calcification of the Fibrous Skeleton of the Base of the Heart and Aortosclerosis (the Cardiovascular Health Study). *Am. J. Cardiol.* **97** (9): 1281–1286.

Barczyk, M., S. Carracedo and D. Gullberg (2010). Integrins. *Cell Tissue Res.* **339** (1): 269–280.

Barnes, J. L. and Y. Gorin (2011). Myofibroblast Differentiation during Fibrosis: Role of Nad(P)H Oxidases. *Kidney Int.* **79** (9): 944–956.

Baronas-Lowell, D., J. L. Lauer-Fields and G. B. Fields (2003). Defining the Roles of Collagen and Collagen-Like Proteins within the Proteome. *J. Liquid Chromatogr. Relat. Technol.* **26** (14): 2225–2254.

Barrett, S. R. H., M. P. F. Sutcliffe, S. Howarth, Z. Y. Li and J. H. Gillard (2009). Experimental Measurement of the Mechanical Properties of Carotid Atherothrombotic Plaque Fibrous Cap. *J. Biomech.* **42** (11): 1650–1655.

Barry, D. (2013). July 31, 2013. http://www.youtube.com/watch?v=yqESR7E4b_8.

Barry, D. M., C. Carpenter, C. Yager, B. Golik, K. J. Barry, H. Shen, et al. (2010). Variation of the Neurofilament Medium Ksp Repeat Sub-Domain across Mammalian Species: Implications for Altering Axonal Structure. *J. Exp. Biol.* **213** (1): 128–136.

Barry, D. M., S. Millecamps, J. P. Julien and M. L. Garcia (2007). New Movements in Neurofilament Transport, Turnover and Disease. *Exp. Cell Res.* **313** (10): 2110–2120.

Bassett, A., S. Cooper, C. Wu and A. Travers (2009). The Folding and Unfolding of Eukaryotic Chromatin. *Curr. Opin. Genet. Dev.* **19** (2): 159–165.

Beel, J. A., D. E. Groswald and M. W. Luttges (1984). Alterations in the Mechanical Properties of Peripheral Nerve Following Crush Injury. *J. Biomech.* **17** (3): 185–193.

Benner, S. A. (2010). Q&A: Life, Synthetic Biology and Risk. *BMC Biol.* **8**: 77.

Bercoff, J., S. Chaffai, M. Tanter, L. Sandrin, S. Catheline, M. Fink, et al. (2003). In vivo Breast Tumor Detection Using Transient Elastography. *Ultrasound Med. Biol.* **29** (10): 1387–1396.

Berg, H. (1999). Problems of Weak Electromagnetic Field Effects in Cell Biology. *Bioelectrochem Bioenerg* **48** (2): 355–360.

Berg, H. C. (1993). *Random Walks in Biology*. Princeton, NJ, Princeton University Press.

Berry, M. F., A. J. Engler, Y. J. Woo, T. J. Pirolli, L. T. Bish, V. Jayasankar, et al. (2006). Mesenchymal Stem Cell Injection after Myocardial Infarction Improves Myocardial Compliance. *Am. J. Phys.-Heart Circ. Physiol.* **290** (6): H2196–H2203.

Berry, R. M. and H. C. Berg (1999). Torque Generated by the Flagellar Motor of Escherichia Coli While Driven Backward. *Biophys. J.* **76** (1 Pt 1): 580–587.

Bershadsky, A. D., O. Y. Ivanova, L. A. Lyass, O. Y. Pletyushkina, J. M. Vasiliev and I. M. Gelfand (1990). Cytoskeletal Reorganizations Responsible for the Phorbol Ester-Induced Formation of Cytoplasmic Processes: Possible Involvement of Intermediate Filaments. *Proc. Natl. Acad. Sci. U. S. A.* **87** (5): 1884–1888.

Binnig, G., C. F. Quate and C. Gerber (1986). Atomic Force Microscope. *Phys. Rev. Lett.* **56** (9): 930–933.

Black, M. M. and R. J. Lasek (1979). Axonal Transport of Actin: Slow Component B Is the Principal Source of Actin for the Axon. *Brain Res.* **171** (3): 401–413.

Black, M. M. and R. J. Lasek (1980). Slow Components of Axonal Transport: Two Cytoskeletal Networks. *J. Cell Biol.* **86** (2): 616–623.

Bloor, M. I. G. and M. J. Wilson (2000). Method for Efficient Shape Parametrization of Fluid Membranes and Vesicles. *Phys. Rev. E* **61** (4): 4218–4229.

Boal, D. H. (2002). *Mechanics of the Cell*. Cambridge, UK; New York, Cambridge University Press.

Bond, J. E., T. Q. Ho, M. A. Selim, C. L. Hunter, E. V. Bowers and H. Levinson (2011). Temporal Spatial Expression and Function of Non-Muscle

Myosin Ii Isoforms Iia and Iib in Scar Remodeling. *Lab Invest.* **91** (4): 499–508.

Bonow, R. O., B. A. Carabello, K. Chatterjee, A. C. de Leon, Jr., D. P. Faxon, M. D. Freed, et al. (2008). 2008 Focused Update Incorporated into the Acc/Aha 2006 Guidelines for the Management of Patients with Valvular Heart Disease: A Report of the American College of Cardiology/American Heart Association Task Force on Practice Guidelines (Writing Committee to Revise the 1998 Guidelines for the Management of Patients with Valvular Heart Disease): Endorsed by the Society of Cardiovascular Anesthesiologists, Society for Cardiovascular Angiography and Interventions, and Society of Thoracic Surgeons. *Circulation* **118** (15): e523–e661.

Bora, F. W., Jr., S. Richardson and J. Black (1980). The Biomechanical Responses to Tension in a Peripheral Nerve. *J. Hand Surg. Am.* **5** (1): 21–25.

Bouffioux, O., A. Berquand, M. Eeman, M. Paquot, Y. F. Dufrene, R. Brasseur and M. Deleu (2007). Molecular Organization of Surfactin-Phospholipid Monolayers: Effect of Phospholipid Chain Length and Polar Head. *Biochim. Biophys. Acta-Biomembr.* **1768** (7): 1758–1768.

Boyle, G. (2004). *Renewable Energy: Power for a Sustainable Future*, Oxford University Press.

Brady, S. T. and R. J. Lasek (1982). Axonal Transport: A Cell-Biological Method for Studying Proteins That Associate with the Cytoskeleton. *Methods Cell Biol.* **25**(Pt B): 365–398.

Bray, D. (1979). Mechanical Tension Produced by Nerve Cells in Tissue Culture. *J. Cell Sci.* **37**: 391–410.

Bray, D. (1984). Axonal Growth in Response to Experimentally Applied Mechanical Tension. *Dev. Biol.* **102** (2): 379–389.

Bray, D. and D. Gilbert (1981). Cytoskeletal Elements in Neurons. *Annu. Rev. Neurosci.* **4**: 505–523.

Bray, D., C. Thomas and G. Shaw (1978). Growth Cone Formation in Cultures of Sensory Neurons. *Proc. Natl. Acad. Sci. U. S. A.* **75** (10): 5226–5229.

Breitbart, M., L. R. Thompson, C. A. Suttle and M. B. Sullivan (2007). Exploring the Vast Diversity of Marine Viruses. *Oceanography* **20** (2): 135–139.

Bridger, J. M., N. Foeger, I. R. Kill and H. Herrmann (2007). The Nuclear Lamina. Both a Structural Framework and a Platform for Genome Organization. *FEBS J.* **274** (6): 1354–1361.

Bridgman, P. C. (2009). Myosin Motor Proteins in the Cell Biology of Axons and Other Neuronal Compartments. *Results Probl. Cell Differ.* **48**: 91–105.

Brochu, H. and P. Vermette (2008). Young's Moduli of Surface-Bound Liposomes by Atomic Force Microscopy Force Measurements. *Langmuir* **24** (5): 2009–2014.

Broers, J. L., E. A. Peeters, H. J. Kuijpers, J. Endert, C. V. Bouten, C. W. Oomens, et al. (2004). Decreased Mechanical Stiffness in Lmna-/- Cells Is Caused by Defective Nucleo-Cytoskeletal Integrity: Implications for the Development of Laminopathies. *Human Mol. Genet.* **13** (21): 2567–2580.

Brown, R., R. Pedowitz, B. Rydevik, S. Woo, A. Hargens, J. Massie, et al. (1993). Effects of Acute Graded Strain on Efferent Conduction Properties in the Rabbit Tibial Nerve. *Clin. Orthop. Relat. Res.* **296**: 288–294.

Buehler, M. J. and T. Ackbarow (2008). Nanomechanical Strength Mechanisms of Hierarchical Biological Materials and Tissues. *Comput. Methods Biomechan. Biomed. Eng.* **11** (6): 595–607.

Bueno, F. R. and S. B. Shah (2008). Implications of Tensile Loading for the Tissue Engineering of Nerves. *Tissue Eng. Part B Rev.* **14** (3): 219–233.

Buki, A. and J. T. Povlishock (2006). All Roads Lead to Disconnection?—Traumatic Axonal Injury Revisited. *Acta Neurochir.* (*Wien*) **148** (2): 181–193; discussion 193–184.

Burridge, K. and K. Fath (1989). Focal Contacts: Transmembrane Links between the Extracellular Matrix and the Cytoskeleton. *Bioessays* **10** (4): 104–108.

Burridge, K., L. Molony and T. Kelly (1987). Adhesion Plaques: Sites of Transmembrane Interaction between the Extracellular Matrix and the Actin Cytoskeleton. *J. Cell Sci. Suppl* **8**: 211–229.

Burton, P. R. (1987). Microtubules of Frog Olfactory Axons: Their Length and Number/Axon. *Brain Res.* **409** (1): 71–78.

Bussiek, M., G. Muller, W. Waldeck, S. Diekmann and J. Langowski (2007). Organisation of Nucleosomal Arrays Reconstituted with Repetitive African Green Monkey Alpha-Satellite DNA as Analysed by Atomic Force Microscopy. *Eur. Biophys. J.* **37** (1): 81–93.

Bustamante, C., J. C. Macosko and G. J. Wuite (2000). Grabbing the Cat by the Tail: Manipulating Molecules One by One. *Nat. Rev. Mol. Cell Biol.* **1** (2): 130–136.

Butcher, J. T., A. M. Penrod, A. J. Garcia and R. M. Nerem (2004). Unique Morphology and Focal Adhesion Development of Valvular Endothelial

Cells in Static and Fluid Flow Environments. *Arterioscler. Thromb. Vasc. Biol.* **24** (8): 1429–1434.

Butcher, J. T., C. A. Simmons and J. N. Warnock (2008). Mechanobiology of the Aortic Heart Valve. *J. Heart Valve Dis.* **17** (1): 62–73.

Butcher, J. T., S. Tressel, T. Johnson, D. Turner, G. Sorescu, H. Jo and R. M. Nerem (2006). Transcriptional Profiles of Valvular and Vascular Endothelial Cells Reveal Phenotypic Differences: Influence of Shear Stress. *Arterioscler. Thromb. Vasc. Biol.* **26** (1): 69–77.

Caille, N., O. Thoumine, Y. Tardy and J. J. Meister (2002). Contribution of the Nucleus to the Mechanical Properties of Endothelial Cells. *J. Biomech.* **35** (2): 177–187.

Calhoun, K. A. and J. R. Swartz (2005). An Economical Method for Cell-Free Protein Synthesis Using Glucose and Nucleoside Monophosphates. *Biotechnol. Prog.* **21** (4): 1146–1153.

Califano, J. P. and C. A. Reinhart-King (2008). A Balance of Substrate Mechanics and Matrix Chemistry Regulates Endothelial Cell Network Assembly. *Cell. Mol. Bioeng.* **1** (2–3): 122–132.

Califano, J. P. and C. A. Reinhart-King (2010). Substrate Stiffness and Cell Area Predict Cellular Traction Stresses in Single Cells and Cells in Contact. *Cell. Mol. Bioeng.* **3** (1): 68–75.

Carlson, M. A., M. T. Longaker and J. S. Thompson (2003). Wound Splinting Regulates Granulation Tissue Survival. *J. Surg. Res.* **110** (1): 304–309.

Carter, S. B. (1967). Haptotoxis and the Mechanism of Cell Motility. *Nature* **213**: 256–260.

Caswell, C. C., M. Barczyk, D. R. Keene, E. Lukomska, D. E. Gullberg and S. Lukomski (2008). Identification of the First Prokaryotic Collagen Sequence Motif That Mediates Binding to Human Collagen Receptors, Integrins Alpha2beta1 and Alpha11beta1. *J. Biol. Chem.* **283** (52): 36168–36175.

Cavalier-Smith, T. (2001). Obcells as Proto-Organisms: Membrane Heredity, Lithophosphorylation, and the Origins of the Genetic Code, the First Cells, and Photosynthesis. *J. Mol. Evol.* **53** (4–5): 555–595.

Cevc, G., A. Schatzlein and H. Richardsen (2002). Ultradeformable Lipid Vesicles Can Penetrate the Skin and Other Semi-Permeable Barriers Unfragmented. Evidence from Double Label Clsm Experiments and Direct Size Measurements. *Biochim. Biophys. Acta* **1564** (1): 21–30.

Chada, S., P. Lamoureux, R. E. Buxbaum and S. R. Heidemann (1997). Cytomechanics of Neurite Outgrowth from Chick Brain Neurons. *J. Cell Sci.* **110** (Pt 10): 1179–1186.

Challacombe, J. F., D. M. Snow and P. C. Letourneau (1996). Role of the Cytoskeleton in Growth Cone Motility and Axonal Elongation. *Semin. Neurosci.* **8**: 67–80.

Chan, K. L., K. Teo, J. Tam and J. G. Dumesnil (2007). Rationale, Design, and Baseline Characteristics of a Randomized Trial to Assess the Effect of Cholesterol Lowering on the Progression of Aortic Stenosis: The Aortic Stenosis Progression Observation: Measuring Effects of Rosuvastatin (Astronomer) Trial. *Am. Heart J.* **153** (6): 925–931.

Chan, M. W., P. D. Arora, P. Bozavikov and C. A. McCulloch (2009). Fak, Pip5kigamma and Gelsolin Cooperatively Mediate Force-Induced Expression of Alpha-Smooth Muscle Actin. *J. Cell Sci.* **122** (Pt 15): 2769–2781.

Chan, M. W., F. Chaudary, W. Lee, J. W. Copeland and C. A. McCulloch (2010). Force-Induced Myofibroblast Differentiation through Collagen Receptors Is Dependent on Mammalian Diaphanous (Mdia). *J. Biol. Chem.* **285** (12): 9273–9281.

Chan, W. K., J. T. Yabe, A. F. Pimenta, D. Ortiz and T. B. Shea (2003). Growth Cones Contain a Dynamic Population of Neurofilament Subunits. *Cell Motil Cytoskeleton* **54** (3): 195–207.

Chancellor, T. J., J. Lee, C. K. Thodeti and T. Lele (2010). Actomyosin Tension Exerted on the Nucleus through Nesprin-1 Connections Influences Endothelial Cell Adhesion, Migration, and Cyclic Strain-Induced Reorientation. *Biophys. J.* **99** (1): 115–123.

Charras, G. T., B. A. Williams, S. M. Sims and M. A. Horton (2004). Estimating the Sensitivity of Mechanosensitive Ion Channels to Membrane Strain and Tension. *Biophys. J.* **87** (4): 2870–2884.

Chary, K. V., R. V. Hosur, G. Govil, Z. K. Tan and H. T. Miles (1987). Novel Solution Conformation of DNA Observed in D(Gaattcgaattc) by Two-Dimensional Nmr Spectroscopy. *Biochemistry* **26** (5): 1315–1322.

Chen, C. S., M. Mrksich, S. Huang, G. M. Whitesides and D. E. Ingber (1997). Geometric Control of Cell Life and Death. *Science* **276** (5317): 1425–1428.

Chen, J. H., W. L. Chen, K. L. Sider, C. Y. Yip and C. A. Simmons (2011). β-Catenin Mediates Mechanically Regulated, Transforming Growth Factor-β1-Induced Myofibroblast Differentiation of Aortic Valve Interstitial Cells. *Arterioscler. Thromb. Vasc. Biol.* 590–597.

Chen, J. H., C. Y. Yip, E. D. Sone and C. A. Simmons (2009). Identification and Characterization of Aortic Valve Mesenchymal Progenitor Cells with Robust Osteogenic Calcification Potential. *Am. J. Pathol.* **174** (3): 1109–1119.

Chen, X., A. Kis, A. Zettl and C. R. Bertozzi (2007). A Cell Nanoinjector Based on Carbon Nanotubes. *Proc. Natl. Acad. Sci. U. S. A.* **104** (20): 8218–8222.

Chen, X. B., L. S. Niu and H. J. Shi (2008). Modeling the Phase Separation in Binary Lipid Membrane under Externally Imposed Oscillatory Shear Flow. *Colloids Surf. B Biointerfaces* **65** (2): 203–212.

Cheng, C. M., P. R. LeDuc and Y. W. Lin (2011). Localized Bimodal Response of Neurite Extensions and Structural Proteins in Dorsal-Root Ganglion Neurons with Controlled Polydimethylsiloxane Substrate Stiffness. *J. Biomech.* **44** (5): 856–862.

Chetta, J., C. Kye and S. B. Shah (2010). Cytoskeletal Dynamics in Response to Tensile Loading of Mammalian Axons. *Cytoskeleton (Hoboken)* **67** (10): 650–665.

Chevalier-Larsen, E. and E. L. Holzbaur (2006). Axonal Transport and Neurodegenerative Disease. *Biochim. Biophys. Acta* **1762** (11–12): 1094–1108.

Ching, G. Y. and R. K. Liem (1993). Assembly of Type IV Neuronal Intermediate Filaments in Nonneuronal Cells in the Absence of Preexisting Cytoplasmic Intermediate Filaments. *J. Cell Biol.* **122** (6): 1323–1335.

Choi, S., W. Wang, A. J. Ribeiro, A. Kalinowski, S. Q. Gregg, P. L. Opresko, et al. (2011). Computational Image Analysis of Nuclear Morphology Associated with Various Nuclear-Specific Aging Disorders. *Nucleus* **2** (6): 570–579.

Chopra, A., E. Tabdanov, H. Patel, P. A. Janmey and J. Y. Kresh (2011). Cardiac Myocyte Remodeling Mediated by N-Cadherin-Dependent Mechanosensing. *Am. J. Physiol.-Heart Circ. Physiol.* **300** (4): H1252–H1266.

Choquet, D., D. P. Felsenfeld and M. P. Sheetz (1997). Extracellular Matrix Rigidity Causes Strengthening of Integrin-Cytoskeleton Linkages. *Cell* **88** (1): 39–48.

Chretien, D., S. D. Fuller and E. Karsenti (1995). Structure of Growing Microtubule Ends: Two-Dimensional Sheets Close into Tubes at Variable Rates. *J. Cell Biol.* **129** (5): 1311–1328.

Chuong, C. J. and Y. C. Fung (1986). On Residual Stresses in Arteries. *J. Biomech. Eng.* **108** (2): 189–192.

Clark-Greuel, J. N., J. M. Connolly, E. Sorichillo, N. R. Narula, H. S. Rapoport, E. R. Mohler, III, et al. (2007). Transforming Growth Factor-Beta1 Mechanisms in Aortic Valve Calcification: Increased Alkaline Phosphatase and Related Events. *Ann. Thorac. Surg.* **83** (3): 946–953.

Cleveland, D. W. (1996). Neuronal Growth and Death: Order and Disorder in the Axoplasm. *Cell* **84** (5): 663–666.

Compiani, M. and E. Capriotti (2013). Computational and Theoretical Methods for Protein Folding. *Biochemistry* **52** (48): 8601–8624.

Cooke, I. R. and M. Deserno (2006). Coupling between Lipid Shape and Membrane Curvature. *Biophys. J.* **91** (2): 487–495.

Coppieters, M. W., M. Van de Velde and K. H. Stappaerts (2002). Positioning in Anesthesiology: Toward a Better Understanding of Stretch-Induced Perioperative Neuropathies. *Anesthesiology* **97** (1): 75–81.

Coppin, C. M., J. T. Finer, J. A. Spudich and R. D. Vale (1996). Detection of Sub-8-Nm Movements of Kinesin by High-Resolution Optical-Trap Microscopy *Proc. Natl. Acad. Sci U. S. A.* **93** (5): 1913–1917.

Cullen, D. K. and M. C. LaPlaca (2006). Neuronal Response to High Rate Shear Deformation Depends on Heterogeneity of the Local Strain Field. *J. Neurotrauma* **23** (9): 1304–1319.

Cushing, M. C., J. T. Liao and K. S. Anseth (2005). Activation of Valvular Interstitial Cells Is Mediated by Transforming Growth Factor-Beta1 Interactions with Matrix Molecules. *Matrix Biol.* **24** (6): 428–437.

Cushing, M. C., P. D. Mariner, J. T. Liao, E. A. Sims and K. S. Anseth (2008). Fibroblast Growth Factor Represses Smad-Mediated Myofibroblast Activation in Aortic Valvular Interstitial Cells. *Faseb J.* **22** (6): 1769–1777.

Dahl, K. N., A. J. Engler, J. D. Pajerowski and D. E. Discher (2005). Power-Law Rheology of Isolated Nuclei with Deformation Mapping of Nuclear Substructures. *Biophys. J.* **89** (4): 2855–2864.

Dahl, K. N., S. M. Kahn, K. L. Wilson and D. E. Discher (2004). The Nuclear Envelope Lamina Network Has Elasticity and a Compressibility Limit Suggestive of a Molecular Shock Absorber. *Journal of Cell Science* **117** (Pt 20): 4779–4786.

Dahl, K. N. and A. Kalinowski (2011). Nucleoskeleton Mechanics at a Glance. *J. Cell Sci.* **124** (Pt 5): 675–678.

Dahl, K. N., A. Kalinowski and K. Pekkan (2010). Mechanobiology and the Microcirculation: Cellular, Nuclear and Fluid Mechanics. *Microcirculation* **17** (3): 179–191.

Dahl, K. N., P. Scaffidi, M. F. Islam, A. G. Yodh, K. L. Wilson and T. Misteli (2006). Distinct Structural and Mechanical Properties of the Nuclear Lamina in Hutchinson-Gilford Progeria Syndrome. *Proc. Natl. Acad. Sci. U. S. A.* **103** (27): 10271–10276.

Dai, J. and M. P. Sheetz (1995). Mechanical Properties of Neuronal Growth Cone Membranes Studied by Tether Formation with Laser Optical Tweezers. *Biophys. J.* **68** (3): 988–996.

Daigle, N., J. Beaudouin, L. Hartnell, G. Imreh, E. Hallberg, J. Lippincott-Schwartz and J. Ellenberg (2001). Nuclear Pore Complexes Form Immobile Networks and Have a Very Low Turnover in Live Mammalian Cells. *J. Cell Biol.* **154** (1): 71–84.

Dansky, L. and E. M. Johnson (1986). Effects of Agents That Perturb Collagen and Glycoprotein on Hydra Embryos. *Teratology* **33** (3): C62–C62.

Davies, P. F., A. G. Passerini and C. A. Simmons (2004). Aortic Valve: Turning over a New Leaf(Let) in Endothelial Phenotypic Heterogeneity. *Arterioscler. Thromb. Vasc. Biol.* **24** (8): 1331–1333.

Dawkins, R. (1976). *The Selfish Gene*. New York, Oxford University Press.

Dawkins, R. (2004). *The Ancestor's Tale: A Pilgrimage to the Dawn of Evolution*. Boston, Houghton Mifflin.

Dawkins, R. (2008). *The Oxford Book of Modern Science Writing*. Oxford, Oxford University Press.

de Lanerolle, P., T. Johnson and W. A. Hofmann (2005). Actin and Myosin I in the Nucleus: What Next? *Nat. Struct. Mol. Biol.* **12** (9): 742–746.

de Waegh, S. M., V. M. Lee and S. T. Brady (1992). Local Modulation of Neurofilament Phosphorylation, Axonal Caliber, and Slow Axonal Transport by Myelinating Schwann Cells. *Cell* **68** (3): 451–463.

Deck, J. D. (1986). Endothelial Cell Orientation on Aortic Valve Leaflets. *Cardiovasc. Res.* **20** (10): 760–767.

Deguchi, S., K. Maeda, T. Ohashi and M. Sato (2005). Flow-Induced Hardening of Endothelial Nucleus as an Intracellular Stress-Bearing Organelle. *J. Biomech.* **38** (9): 1751–1759.

Delbarre, E., M. Tramier, M. Coppey-Moisan, C. Gaillard, J. C. Courvalin and B. Buendia (2006). The Truncated Prelamin a in Hutchinson-Gilford Progeria Syndrome Alters Segregation of a-Type and B-Type Lamin Homopolymers. *Human Mol. Genet.* **15** (7): 1113–1122.

Dembo, M. and Y. L. Wang (1999). Stresses at the Cell-to-Substrate Interface during Locomotion of Fibroblasts. *Biophys. J.* **76** (4): 2307–2316.

Dennerll, T. J., H. C. Joshi, V. L. Steel, R. E. Buxbaum and S. R. Heidemann (1988). Tension and Compression in the Cytoskeleton of Pc-12 Neurites. Ii: Quantitative Measurements. *J. Cell Biol.* **107** (2): 665–674.

Desmouliere, A., M. Redard, I. Darby and G. Gabbiani (1995). Apoptosis Mediates the Decrease in Cellularity during the Transition between Granulation Tissue and Scar. *Am. J. Pathol.* **146** (1): 56–66.

Dhe-Paganon, S., E. D. Werner, Y. I. Chi and S. E. Shoelson (2002). Structure of the Globular Tail of Nuclear Lamin. *J. Biol. Chem.* **277** (20): 17381–17384.

Di Lullo, G. A., S. M. Sweeney, J. Korkko, L. Ala-Kokko and J. D. San Antonio (2002). Mapping the Ligand-Binding Sites and Disease-Associated Mutations on the Most Abundant Protein in the Human, Type I Collagen. *J. Biol. Chem.* **277** (6): 4223–4231.

Dickerson, R. E. (1989). Definitions and Nomenclature of Nucleic Acid Structure Components. *Nucleic Acids Res.* **17** (5): 1797–1803.

Dill, K. A. and S. Bromberg (2011). *Molecular Driving Forces: Statistical Thermodynamics in Biology, Chemistry, Physics, and Nanoscience.* London; New York, Garland Science.

Dill, K. A., S. Bromberg, K. Yue, K. M. Fiebig, D. P. Yee, P. D. Thomas and H. S. Chan (1995). Principles of Protein Folding—a Perspective from Simple Exact Models. *Protein Sci.* **4** (4): 561–602.

Dimilla, P. A., K. Barbee and D. A. Lauffenburger (1991). Mathematical-Model for the Effects of Adhesion and Mechanics on Cell-Migration Speed. *Biophys. J.* **60** (1): 15–37.

Dimilla, P. A., J. A. Stone, J. A. Quinn, S. M. Albelda and D. A. Lauffenburger (1993). Maximal Migration of Human Smooth-Muscle Cells on Fibronectin and Type–IV Collagen Occurs at an Intermediate Attachment Strength. *J. Cell Biol.* **122** (3): 729–737.

Discher, D. E., P. Janmey and Y. L. Wang (2005). Tissue Cells Feel and Respond to the Stiffness of Their Substrate. *Science* **310** (5751): 1139–1143.

Dokukin, M. E., N. V. Guz and I. Sokolov (2013). Quantitative Study of the Elastic Modulus of Loosely Attached Cells in Afm Indentation Experiments. *Biophys. J.* **104** (10): 2123–2131.

Dorner, D., J. Gotzmann and R. Foisner (2007). Nucleoplasmic Lamins and Their Interaction Partners, Lap2alpha, Rb, and Baf, in Transcriptional Regulation. *FEBS J.* **274** (6): 1362–1373.

Duband, J. L., S. Dufour, S. S. Yamada, K. M. Yamada and J. P. Thiery (1991). Neural Crest Cell Locomotion Induced by Antibodies to Beta-1 Integrins—a Tool for Studying the Roles of Substratum Molecular Avidity and Density in Migration. *J. Cell Sci.* **98**: 517–532.

Dugina, V., L. Fontao, C. Chaponnier, J. Vasiliev and G. Gabbiani (2001). Focal Adhesion Features during Myofibroblastic Differentiation Are Controlled by Intracellular and Extracellular Factors. *J. Cell Sci.* **114** (Pt 18): 3285–3296.

Duncan, J. E. and L. S. Goldstein (2006). The Genetics of Axonal Transport and Axonal Transport Disorders. *PLoS Genet* **2** (9): e124.

Dyson, F. (2007). Our Biotech Future. *N. York Rev. Books* **54** (12): 4–8.

Eastwood, M., V. C. Mudera, D. A. McGrouther and R. A. Brown (1998). Effect of Precise Mechanical Loading on Fibroblast Populated Collagen Lattices: Morphological Changes. *Cell Motil. Cytoskeleton* **40** (1): 13–21.

EH, E. and P. R (1984). X-Ray Diffraction Evidence That Actin Is a 100 Angstrom Filament. *Nature* **307**: 56–58.

El-Hamamsy, I., K. Balachandran, M. H. Yacoub, L. M. Stevens, P. Sarathchandra, P. M. Taylor, et al. (2009). Endothelium-Dependent Regulation of the Mechanical Properties of Aortic Valve Cusps. *J. Am. Coll. Cardiol.* **53** (16): 1448–1455.

Elston, T., H. Wang and G. Oster (1998). Energy Transduction in Atp Synthase. *Nature* **391** (6666): 510–513.

Engler, A., L. Bacakova, C. Newman, A. Hategan, M. Griffin and D. Discher (2004a). Substrate Compliance Versus Ligand Density in Cell on Gel Responses. *Biophys. J.* **86** (1): 617–628.

Engler, A. J., C. Carag-Krieger, C. P. Johnson, M. Raab, H. Y. Tang, D. W. Speicher, et al. (2008). Embryonic Cardiomyocytes Beat Best on a Matrix with Heart-Like Elasticity: Scar-Like Rigidity Inhibits Beating. *J. Cell Sci.* **121** (22): 3794–3802.

Engler, A. J., M. A. Griffin, S. Sen, C. G. Bonnetnann, H. L. Sweeney and D. E. Discher (2004b). Myotubes Differentiate Optimally on Substrates with Tissue-Like Stiffness: Pathological Implications for Soft or Stiff Microenvironments. *J. Cell Biol.* **166** (6): 877–887.

Engler, A. J., S. Sen, H. L. Sweeney and D. E. Discher (2006). Matrix Elasticity Directs Stem Cell Lineage Specification. *Cell* **126** (4): 677–689.

Erickson, B. N., H. H. Williams, S. S. Bernstein, I. Avrin, R. L. Jones and I. G. Macy (1937). The Lipid Distribution of Posthemolytic Residue or Stroma of Erythrocytes. *J. Biol. Chem.* **115**: 515–528.

Eriksen, H. A., J. Satta, J. Risteli, M. Veijola, P. Vare and Y. Soini (2006). Type I and Type III Collagen Synthesis and Composition in the Valve Matrix in Aortic Valve Stenosis. *Atherosclerosis* **189** (1): 91–98.

Erler, J. T., K. L. Bennewith, T. R. Cox, G. Lang, D. Bird, A. Koong, et al. (2009). Hypoxia-Induced Lysyl Oxidase Is a Critical Mediator of Bone Marrow Cell Recruitment to Form the Premetastatic Niche. *Cancer Cell.* **15** (1): 35–44.

Evans, E. and W. Rawicz (1990). Entropy-Driven Tension and Bending Elasticity in Condensed-Fluid Membranes. *Phys. Rev. Lett.* **64** (17): 2094–2097.

Everts, V. and W. Beertsen (1996). Phagocytosis and Intracellular Digestion of Collagen, Its Role in Turnover and Remodelling. *Histochem. J.* **28**: 229–245.

Falnikar, A., S. Tole and P. W. Baas (2011). Kinesin-5, a Mitotic Microtubule-Associated Motor Protein, Modulates Neuronal Migration. *Mol. Biol. Cell.* **22** (9): 1561–1574.

Farivar, R. S., L. H. Cohn, E. G. Soltesz, T. Mihaljevic, J. D. Rawn and J. G. Byrne (2003). Transcriptional Profiling and Growth Kinetics of Endothelium Reveals Differences between Cells Derived from Porcine Aorta Versus Aortic Valve. *Eur. J. Cardiothorac. Surg.* **24** (4): 527–534.

Feng, F. and W. S. Klug (2006). Finite Element Modeling of Lipid Bilayer Membranes. *J. Comput. Phys.* **220** (1): 394–408.

Fernandez, P. and P. A. Pullarkat (2010). The Role of the Cytoskeleton in Volume Regulation and Beading Transitions in Pc12 Neurites. *Biophys. J.* **99** (11): 3571–3579.

Feynman, R. P., R. B. Leighton and M. L. Sands (1963). *The Feynman Lectures on Physics*. Reading, Mass. Addison-Wesley Pub. Co.

Figueroa, D. S., S. F. Kemeny and A. M. Clyne (2011). Glycated Collagen Impairs Endothelial Cell Response to Cyclic Stretch. *Cell. Mol. Bioeng.* **4** (2): 220–230.

Finer, J. T., R. M. Simmons, and J. A. Spudich (1994). Single Myosin Molecule Mechanics: Piconewton Forces and Nanometre Steps. *Nature* **368**(6467): 113–119.

Finlay, B. J. (2002). Global Dispersal of Free-Living Microbial Eukaryote Species. *Science* **296** (5570): 1061–1063.

Flanagan, L. A., Y. E. Ju, B. Marg, M. Osterfield and P. A. Janmey (2002). Neurite Branching on Deformable Substrates. *Neuroreport* **13** (18): 2411–2415.

Flynn, K. C., C. W. Pak A. E. Shaw, F. Bradke and J. R. Bamburg (2009). Growth Cone-Like Waves Transport Actin and Promote Axonogenesis and Neurite Branching. *Dev. Neurobiol.* **69** (12): 761–779.

Foisner, R. and L. Gerace (1993). Integral Membrane Proteins of the Nuclear Envelope Interact with Lamins and Chromosomes, and Binding Is Modulated by Mitotic Phosphorylation. *Cell* **73** (7): 1267–1279.

Foloppe, N., L. Nilsson and A. D. MacKerell, Jr. (2001). Ab Initio Conformational Analysis of Nucleic Acid Components: Intrinsic

Energetic Contributions to Nucleic Acid Structure and Dynamics. *Biopolymers* **61** (1): 61–76.

Fondard, O., D. Detaint, B. Iung, C. Choqueux, H. dle-Biassette, M. Jarraya, et al. (2005). Extracellular Matrix Remodelling in Human Aortic Valve Disease: The Role of Matrix Metalloproteinases and Their Tissue Inhibitors. *Eur. Heart J.* **26** (13): 1333–1341.

Fong, L. G., J. K. Ng, J. Lammerding, T. A. Vickers, M. Meta, N. Cote, et al. (2006). Prelamin a and Lamin a Appear to Be Dispensable in the Nuclear Lamina. *J. Clin. Invest.* **116** (3): 743–752.

Fortenberry, C., E. A. Bowman, W. Proffitt, B. Dorr, S. Combs, J. Harp, et al. (2011). Exploring Symmetry as an Avenue to the Computational Design of Large Protein Domains. *J. Am. Chem. Soc.* **133** (45): 18026–18029.

Forterre, P. (2010). Defining Life: The Virus Viewpoint. *Orig. Life Evol. Biosph.* **40** (2): 151–160.

Frappier, T., F. Stetzkowski-Marden and L. A. Pradel (1991). Interaction Domains of Neurofilament Light Chain and Brain Spectrin. *Biochem. J.* **275** (Pt 2): 521–527.

Freeman, R. V. and C. M. Otto (2005). Spectrum of Calcific Aortic Valve Disease: Pathogenesis, Disease Progression, and Treatment Strategies. *Circulation* **111** (24): 3316–3326.

Fuchs, C., E. Glasgow, P. F. Hitchcock and N. Schechter (1994). Plasticin, a Newly Identified Neurofilament Protein, Is Preferentially Expressed in Young Retinal Ganglion Cells of Adult Goldfish. *J. Comp. Neurol.* **350** (3): 452–462.

Fuhrman, J. A., K. McCallum and A. A. Davis (1993). Phylogenetic Diversity of Subsurface Marine Microbial Communities from the Atlantic and Pacific Oceans. *Appl. Environ. Microbiol.* **59** (5): 1294–1302.

Fujii, T., M. Watanabe, Y. Ogoma, Y. Kondo and T. Arai (1993). Microtubule-Associated Proteins, Map 1a and Map 1b, Interact with F-Actin *in vitro*. *J. Biochem.* **114** (6): 827–829.

Fung, Y. C. (1993). *Biomechanics: Mechanical Properties of Living Tissues.* New York, Springer-Verlag.

Futerman, A. H. and G. A. Banker (1996). The Economics of Neurite Outgrowth—the Addition of New Membrane to Growing Axons. *Trends Neurosci.* **19** (4): 144–149.

Fygenson, D. K., M. Elbaum, B. Shraiman and A. Libchaber (1997). Microtubules and Vesicles under Controlled Tension. *Phys. Rev. E.* **55** (1): 850–859.

Gabbiani, G., C. Chaponnier and I. Huttner (1978). Cytoplasmic Filaments and Gap Junctions in Epithelial Cells and Myofibroblasts during Wound Healing. *J. Cell Biol.* **76** (3): 561–568.

Gabbiani, G., G. B. Ryan and G. Majne (1971). Presence of Modified Fibroblasts in Granulation Tissue and Their Possible Role in Wound Contraction. *Experientia* **27** (5): 549–550.

Gagliano, J., M. Walb, B. Blaker, J. C. Macosko and G. Holzwarth (2010). Kinesin Velocity Increases with the Number of Motors Pulling against Viscoelastic Drag. *Eur. Biophys. J.* **39** (5): 801–813.

Galbraith, C. G., K. M. Yamada and J. A. Galbraith (2007). Polymerizing Actin Fibers Position Integrins Primed to Probe for Adhesion Sites. *Science* **315** (5814): 992–995.

Galilei, G. (1991). *Dialogues Concerning Two New Sciences.* Buffalo, N. Y., Prometheus Books.

Gallant, N. D., K. E. Michael and A. J. Garcia (2005). Cell Adhesion Strengthening: Contributions of Adhesive Area, Integrin Binding, and Focal Adhesion Assembly. *Mol. Biol. Cell.* **16** (9): 4329–4340.

Garcia, M. L., C. S. Lobsiger, S. B. Shah, T. J. Deerinck, J. Crum, D. Young, et al. (2003). Nf-M Is an Essential Target for the Myelin-Directed "Outside-in" Signaling Cascade That Mediates Radial Axonal Growth. *J. Cell Biol.* **163** (5): 1011–1020.

Garcia, M. L., M. V. Rao, J. Fujimoto, V. B. Garcia, S. B. Shah, J. Crum, et al. (2009). Phosphorylation of Highly Conserved Neurofilament Medium Ksp Repeats Is Not Required for Myelin-Dependent Radial Axonal Growth. *J. Neurosci.* **29** (5): 1277–1284.

Gardner, M. K., B. D. Charlebois, I. M. Janosi, J. Howard, A. J. Hunt and D. J. Odde (2011). Rapid Microtubule Self-Assembly Kinetics. *Cell* **146** (4): 582–592.

Gaudet, C., W. A. Marganski, S. Kim, C. T. Brown, V. Gunderia, M. Dembo and J. Y. Wong (2003). Influence of Type I Collagen Surface Density on Fibroblast Spreading, Motility, and Contractility. *Biophys. J.* **85** (5): 3329–3335.

Geiger, B., A. Bershadsky, R. Pankov and K. M. Yamada (2001). Transmembrane Crosstalk between the Extracellular Matrix—Cytoskeleton Crosstalk. *Nat. Rev. Mol. Cell Biol.* **2** (11): 793–805.

Genes, N. G., J. A. Rowley, D. J. Mooney and L. J. Bonassar (2004). Effect of Substrate Mechanics on Chondrocyte Adhesion to Modified Alginate Surfaces. *Arch. Biochem. Biophys.* **422** (2): 161–167.

Gennerich, A., A. P. Carter, S. L. Reck-Peterson and R. D. Vale (2007). Force-Induced Bidirectional Stepping of Cytoplasmic Dynein. *Cell* **131** (5): 952–965.

Georgatos, S. D., Y. Markaki, A. Christogianni and A. S. Politou (2009). Chromatin Remodeling during Mitosis: A Structure-Based Code? *Front. Biosci.* **14**: 2017–2027.

Georgeu, G. A., E. T. Walbeehm, R. Tillett, A. Afoke, R. A. Brown and J. B. Phillips (2005). Investigating the Mechanical Shear-Plane between Core and Sheath Elements of Peripheral Nerves. *Cell Tissue Res.* **320** (2): 229–234.

Gershon, N. D., K. R. Porter and B. L. Trus (1985). The Cytoplasmic Matrix: Its Volume and Surface Area and the Diffusion of Molecules through It. *Proc. Natl. Acad. Sci. U. S. A.* **82** (15): 5030–5034.

Ghosh, K., Z. Pan, E. Guan, S. R. Ge, Y. J. Liu, T. Nakamura, et al. (2007). Cell Adaptation to a Physiologically Relevant Ecm Mimic with Different Viscoelastic Properties. *Biomaterials* **28** (4): 671–679.

Gieni, R. S. and M. J. Hendzel (2009). Actin Dynamics and Functions in the Interphase Nucleus: Moving toward an Understanding of Nuclear Polymeric Actin. *Biochem. Cell Biol.-Biochim. Biol. Cell.* **87** (1): 283–306.

Gilbert, K. K., J. M. Brismee, D. L. Collins, C. R. James, R. V. Shah, S. F. Sawyer and P. S. Sizer, Jr. (2007a). 2006 Young Investigator Award Winner: Lumbosacral Nerve Root Displacement and Strain: Part 1. A Novel Measurement Technique during Straight Leg Raise in Unembalmed Cadavers. *Spine* **32** (14): 1513–1520.

Gilbert, K. K., J. M. Brismee, D. L. Collins, C. R. James, R. V. Shah, S. F. Sawyer and P. S. Sizer, Jr. (2007b). 2006 Young Investigator Award Winner: Lumbosacral Nerve Root Displacement and Strain: Part 2. A Comparison of 2 Straight Leg Raise Conditions in Unembalmed Cadavers. *Spine* **32** (14): 1521–1525.

Gilbert, N., S. Boyle, H. Fiegler, K. Woodfine, N. P. Carter and W. A. Bickmore (2004). Chromatin Architecture of the Human Genome: Gene-Rich Domains Are Enriched in Open Chromatin Fibers. *Cell* **118** (5): 555–566.

Gittes, F., B. Mickey, J. Nettleton and J. Howard (1993). Flexural Rigidity of Microtubules and Actin Filaments Measured from Thermal Fluctuations in Shape. *J. Cell Biol.* **120** (4): 923–934.

Goffin, J. M., P. Pittet, G. Csucs, J. W. Lussi, J. J. Meister and B. Hinz (2006). Focal Adhesion Size Controls Tension-Dependent Recruitment of

Alpha-Smooth Muscle Actin to Stress Fibers. *J. Cell Biol.* **172** (2): 259–268.

Goldberg, M. W., I. Huttenlauch, C. J. Hutchison and R. Stick (2008). Filaments Made from A- and B-Type Lamins Differ in Structure and Organization. *J. Cell Sci.* **121** (Pt 2): 215–225.

Goldman, R. D., B. Grin, M. G. Mendez and E. R. Kuczmarski (2008). Intermediate Filaments: Versatile Building Blocks of Cell Structure. *Curr. Opin. Cell Biol.* **20** (1): 28–34.

Goldsmith, W. and J. Plunkett (2004). A Biomechanical Analysis of the Causes of Traumatic Brain Injury in Infants and Children. *Am. J. Forensic Med. Pathol.* **25** (2): 89–100.

Gompper, G. and D. M. Kroll (1995). Driven Transport of Fluid Vesicles through Narrow Pores. *Phys. Rev. E. Stat. Phys. Plasmas. Fluids Relat. Interdiscip Topics* **52** (4): 4198–4208.

Gotzmann, J. and R. Foisner (1999). Lamins and Lamin-Binding Proteins in Functional Chromatin Organization. *Crit. Rev. Eukaryot. Gene Exp.* **9** (3–4): 257–265.

Gradl, G., and S. Sterrer (1995). Mobility of Molecules and Particles within the Cytoplasm of a Living Cell. *Microscopy from Carl Zeiss* (micro special 40-619).

Gray, D. S., J. Tien and C. S. Chen (2003). Repositioning of Cells by Mechanotaxis on Surfaces with Micropatterned Young's Modulus. *J. Biomed. Mater. Res. Part A* **66A** (3): 605–614.

Gross, S. P., M. A. Welte, S. M. Block and E. F. Wieschaus (2002). Coordination of Opposite-Polarity Microtubule Motors. *J. Cell Biol* **156** (4): 715–724.

Gruenbaum, Y., A. Margalit, R. D. Goldman, D. K. Shumaker and K. L. Wilson (2005). The Nuclear Lamina Comes of Age. *Nat. Rev. Mol. Cell Biol.* **6** (1): 21–31.

Guck, J., R. Ananthakrishnan, H. Mahmood, T. J. Moon, C. C. Cunningham and J. Kas (2001). The Optical Stretcher: A Novel Laser Tool to Micromanipulate Cells. *Biophys. J.* **81** (2): 767–784.

Guilak, F. (1995). Compression-Induced Changes in the Shape and Volume of the Chondrocyte Nucleus. *J. Biomech.* **28** (12): 1529–1541.

Guilak, F., J. R. Tedrow and R. Burgkart (2000). Viscoelastic Properties of the Cell Nucleus. *Biochem. Biophys. Res. Commun.* **269** (3): 781–786.

Gundersen, R. W. (1988). Interference Reflection Microscopic Study of Dorsal Root Growth Cones on Different Substrates: Assessment of Growth Cone-Substrate Contacts. *J. Neurosci. Res.* **21** (2–4): 298–306.

Gunn, J. W., S. D. Turner and B. K. Mann (2005). Adhesive and Mechanical Properties of Hydrogels Influence Neurite Extension. *J. Biomed. Mater. Res. A* **72** (1): 91–97.

Gupta, V., J. A. Werdenberg, T. L. Blevins and K. J. Grande-Allen (2007). Synthesis of Glycosaminoglycans in Differently Loaded Regions of Collagen Gels Seeded with Valvular Interstitial Cells. *Tissue Eng.* **13** (1): 41–49.

Gupton, S. L. and C. M. Waterman-Storer (2006). Spatiotemporal Feedback between Actomyosin and Focal-Adhesion Systems Optimizes Rapid Cell Migration. *Cell* **125** (7): 1361–1374.

Haghnia, M., V. Cavalli, S. B. Shah, K. Schimmelpfeng, R. Brusch, G. Yang, et al. (2007). Dynactin Is Required for Coordinated Bidirectional Motility, but Not for Dynein Membrane Attachment. *Mol. Biol. Cell* **18** (6): 2081–2089.

Hamill, O. P. and B. Martinac (2001). Molecular Basis of Mechanotransduction in Living Cells. *Physiol. Rev.* **81** (2): 685–740.

Hamon, L., P. A. Curmi and D. Pastre (2010). High-Resolution Imaging of Microtubules and Cytoskeleton Structures by Atomic Force Microscopy. *Methods Cell Biol.* **95**: 157–174.

Haninec, P. (1986). Undulating Course of Nerve Fibres and Bands of Fontana in Peripheral Nerves of the Rat. *Anat. Embryol.* (*Berl*) **174** (3): 407–411.

Hao, R., R. G. MacDonald, M. Ebadi, J. C. Schmit and R. F. Pfeiffer (1997). Stable Interaction between G-Actin and Neurofilament Light Subunit in Dopaminergic Neurons. *Neurochem. Int.* **31** (6): 825–834.

Harborth, J., S. M. Elbashir, K. Bechert, T. Tuschl and K. Weber (2001). Identification of Essential Genes in Cultured Mammalian Cells Using Small Interfering Rnas. *J. Cell Sci.* **114** (Pt 24): 4557–4565.

Harris, A. K., P. Wild and D. Stopak (1980). Silicone-Rubber Substrata-New Wrinkle in the Study of Cell Locomotion. *Science* **208** (4440): 177–179.

Hasaka, T. P., K. A. Myers and P. W. Baas (2004). Role of Actin Filaments in the Axonal Transport of Microtubules. *J. Neurosci.* **24** (50): 11291–11301.

Hasselbach, W. (1978). The Reversibility of the Sarcoplasmic Calcium Pump. *Biochim. Biophys. Acta* **515** (1): 23–53.

He, Y., F. Francis, K. A. Myers, W. Yu, M. M. Black and P. W. Baas (2005). Role of Cytoplasmic Dynein in the Axonal Transport of Microtubules and Neurofilaments. *J. Cell Biol.* **168** (5): 697–703.

Hegner, M. and W. Grange (2002). Mechanics and Imaging of Single DNA Molecules. *J. Muscle Res. Cell Motil.* **23** (5–6): 367–375.

Heidemann, S. R., H. C. Joshi, A. Schechter, J. R. Fletcher and M. Bothwell (1985). Synergistic Effects of Cyclic Amp and Nerve Growth Factor on Neurite Outgrowth and Microtubule Stability of Pc12 Cells. *J. Cell Biol.* **100** (3): 916–927.

Heitlinger, E., M. Peter, A. Lustig, W. Villiger, E. A. Nigg and U. Aebi (1992). The Role of the Head and Tail Domain in Lamin Structure and Assembly: Analysis of Bacterially Expressed Chicken Lamin a and Truncated B2 Lamins. *J. Struct. Biol.* **108** (1): 74–89.

Heldin, C. H., K. Miyazono and P. ten Dijke (1997). TGF-Beta Signalling from Cell Membrane to Nucleus through Smad Proteins. *Nature* **390** (6659): 465–471.

Hendricks, A. G., E. Perlson, J. L. Ross, H. W. Schroeder, 3rd, M. Tokito and E. L. Holzbaur (2010). Motor Coordination Via a Tug-of-War Mechanism Drives Bidirectional Vesicle Transport. *Curr. Biol.* **20** (8): 697–702.

Henriksen, J. R. and J. H. Ipsen (2004). Measurement of Membrane Elasticity by Micro-Pipette Aspiration. *Eur. Phys. J. E Soft Matter* **14** (2): 149–167.

Herrmann, H., H. Bar, L. Kreplak, S. V. Strelkov and U. Aebi (2007). Intermediate Filaments: From Cell Architecture to Nanomechanics. *Nat. Rev. Mol. Cell Biol.* **8** (7): 562–573.

Herrmann, H., M. Hesse, M. Reichenzeller, U. Aebi and T. M. Magin (2003). Functional Complexity of Intermediate Filament Cytoskeletons: From Structure to Assembly to Gene Ablation. *Intl. Rev. Cytol.* **223**: 83–175.

Hildebrand, J. D., J. M. Taylor and J. T. Parsons (1996). An Sh3 Domain-Containing Gtpase-Activating Protein for Rho and Cdc42 Associates with Focal Adhesion Kinase. *Mol. Cell Biol.* **16** (6): 3169–3178.

Hill, D. B., M. J. Plaza, K. Bonin and G. Holzwarth (2004). Fast Vesicle Transport in Pc12 Neurites: Velocities and Forces. *Eur. Biophys. J.* **33** (7): 623–632.

Hill, T. L. (1981). Microfilament or Microtubule Assembly or Disassembly against a Force. *Proc. Natl. Acad. Sci. U. S. A.* **78** (9): 5613–5617.

Hill, T. L. (1986). Effect of Fluctuating Surface Structure and Free Energy on the Growth of Linear Tubular Aggregates. *Biophys. J.* **49** (5): 1017–1031.

Hill, T. L. (1987). *Linear Aggregation Theory in Cell Biology*. New York, Springer-Verlag.

Hingorani, K. S. and L. M. Gierasch (2014). Comparing Protein Folding in vitro and in vivo: Foldability Meets the Fitness Challenge. *Opin. Struct. Biol.* **24C**: 81–90.

Hinz, B. (2006). Masters and Servants of the Force: The Role of Matrix Adhesions in Myofibroblast Force Perception and Transmission. *Eur. J. Cell Biol.* **85** (3–4): 175–181.

Hinz, B. (2007). Formation and Function of the Myofibroblast during Tissue Repair. *J. Invest. Dermatol.* **127** (3): 526–537.

Hinz, B. (2009). Tissue Stiffness, Latent TGF-Beta1 Activation, and Mechanical Signal Transduction: Implications for the Pathogenesis and Treatment of Fibrosis. *Curr. Rheumatol. Rep.* **11** (2): 120–126.

Hinz, B. (2010). The Myofibroblast: Paradigm for a Mechanically Active Cell. *J. Biomech.* **43** (1): 146–155.

Hinz, B., V. Dugina, C. Ballestrem, B. Wehrle-Haller and C. Chaponnier (2003). Alpha-Smooth Muscle Actin Is Crucial for Focal Adhesion Maturation in Myofibroblasts. *Mol. Biol. Cell* **14** (6): 2508–2519.

Hinz, B., G. Gabbiani and C. Chaponnier (2002). The Nh2-Terminal Peptide of Alpha-Smooth Muscle Actin Inhibits Force Generation by the Myofibroblast in vitro and in vivo. *J. Cell Biol.* **157** (4): 657–663.

Hinz, B., D. Mastrangelo, C. E. Iselin, C. Chaponnier and G. Gabbiani (2001). Mechanical Tension Controls Granulation Tissue Contractile Activity and Myofibroblast Differentiation. *Am. J. Pathol.* **159** (3): 1009–1020.

Hinz, B., S. H. Phan, V. J. Thannickal, A. Galli, M. L. Bochaton-Piallat and G. Gabbiani (2007). The Myofibroblast: One Function, Multiple Origins. *Am. J. Pathol.* **170** (6): 1807–1816.

Hinz, B., P. Pittet, J. Smith-Clerc, C. Chaponnier and J. J. Meister (2004). Myofibroblast Development Is Characterized by Specific Cell-Cell Adherens Junctions. *Mol. Biol. Cell* **15** (9): 4310–4320.

Hirayama, Y. and B. E. Sumpio (2007). Role of Ligand-Specific Integrins in Endothelial Cell Alignment and Elongation Induced by Cyclic Strain. *Endothelium* **14** (6): 275–283.

Hirokawa, N. (1982). Cross-Linker System between Neurofilaments, Microtubules, and Membranous Organelles in Frog Axons Revealed by the Quick-Freeze, Deep-Etching Method. *J. Cell Biol.* **94** (1): 129–142.

Hirokawa, N., M. A. Glicksman and M. B. Willard (1984). Organization of Mammalian Neurofilament Polypeptides within the Neuronal Cytoskeleton. *J. Cell Biol.* **98** (4): 1523–1536.

Hirokawa, N. and S. Takeda (1998). Gene Targeting Studies Begin to Reveal the Function of Neurofilament Proteins. *J. Cell Biol.* **143** (1): 1–4.

Hirokawa, N. and R. Takemura (2004). Molecular Motors in Neuronal Development, Intracellular Transport and Diseases. *Curr. Opin. Neurobiol.* **14** (5): 564–573.

Holaska, J. M., K. L. Wilson and M. Mansharamani (2002). The Nuclear Envelope, Lamins and Nuclear Assembly. *Curr. Opin. Cell Biol.* **14** (3): 357–364.

Holmes, K. C., D. Popp, W. Gebhard and W. Kabsch (1990). Atomic Model of the Actin Filament. *Nature* **347** (6288): 44–49.

Hotani, H., T. Inaba, F. Nomura, S. Takeda, K. Takiguchi, T. J. Itoh, et al. (2003). Mechanical Analyses of Morphological and Topological Transformation of Liposomes. *Biosystems* **71** (1–2): 93–100.

Hotani, H. and H. Miyamoto (1990). Dynamic Features of Microtubules as Visualized by Dark-Field Microscopy. *Adv. Biophys.* **26**: 135–156.

Hotani, H., F. Nomura and Y. Suzuki (1999). Giant Liposomes: From Membrane Dynamics to Cell Morphogenesis. *Curr. Opin. Colloid Interface Sci.* **4** (5): 358–368.

Howard, J. (2001). *Mechanics of Motor Proteins and the Cytoskeleton.* Sunderland, Mass., Sinauer Associates, Publishers.

Howard, J., A. J. Hudspeth and R. D. Vale (1989). Movement of Microtubules by Single Kinesin Molecules. *Nature* **342** (6246): 154–158.

Huber, M. D., T. Guan and L. Gerace (2009). Overlapping Functions of Nuclear Envelope Proteins Net25 (Lem2) and Emerin in Regulation of Extracellular Signal-Regulated Kinase Signaling in Myoblast Differentiation. *Mol. Cell. Biol.* **29** (21): 5718–5728.

Huh, D., Y. S. Torisawa, G. A. Hamilton, H. J. Kim and D. E. Ingber (2012). Microengineered Physiological Biomimicry: Organs-on-Chips. *Lab Chip* **12** (12): 2156–2164.

Huh, J. W., M. A. Franklin, A. G. Widing and R. Raghupathi (2006). Regionally Distinct Patterns of Calpain Activation and Traumatic Axonal Injury Following Contusive Brain Injury in Immature Rats. *Dev. Neurosci.* **28** (4–5): 466–476.

Hulbert, A. J. (2003). Life, Death and Membrane Bilayers. *J. Exp. Biol.* **206** (Pt 14): 2303–2311.

Hunt, A. J. and J. Howard (1993). Kinesin Swivels to Permit Microtubule Movement in Any Direction. *Proc. Natl. Acad. Sci. U. S. A.* **90** (24): 11653–11657.

Huttenlocher, A., M. H. Ginsberg and A. F. Horwitz (1996). Modulation of Cell Migration by Integrin-Mediated Cytoskeletal Linkages and Ligand-Binding Affinity. *J. Cell Biol.* **134** (6): 1551–1562.

Huttenlocher, A., S. P. Palecek, Q. Lu, W. L. Zhang, R. L. Mellgren, D. A. Lauffenburger, et al. (1997). Regulation of Cell Migration by the Calcium-Dependent Protease Calpain. *J. Biol. Chem.* **272** (52): 32719–32722.

Huxley, J. (1993). *Problems of Relative Growth*. Baltimore, Johns Hopkins University Press.

Ichimura, H., T. Shiga, I. Abe, Y. Hara, N. Terui, A. Tsujino and N. Ochiai (2005). Distribution of Sodium Channels during Nerve Elongation in Rat Peripheral Nerve. *J. Orthop. Sci.* **10** (2): 214–220.

Ikeda, K., K. Tomita and S. Tanaka (2000). Experimental Study of Peripheral Nerve Injury during Gradual Limb Elongation. *Hand Surg.* **5** (1): 41–47.

Ingber, D. E. (1990). Fibronectin Controls Capillary Endothelial Cell Growth by Modulating Cell Shape. *Proc. Natl. Acad. Sci. U. S. A.* **87** (9): 3579–3583.

Ingber, D. E. (1993). Cellular Tensegrity: Defining New Rules of Biological Design That Govern the Cytoskeleton. *J. Cell Sci.* **104** (Pt 3): 613–627.

Ingber, D. E. (2003a). Mechanobiology and Diseases of Mechanotransduction. *Ann. Med.* **35** (8): 564–577.

Ingber, D. E. (2003b). Tensegrity I. Cell Structure and Hierarchical Systems Biology. *J. Cell Sci.* **116** (Pt 7): 1157–1173.

Jaalouk, D. E. and J. Lammerding (2009). Mechanotransduction Gone Awry. *Nat. Rev. Mol. Cell Biol.* **10** (1): 63–73.

Jacomy, H., Q. Zhu, S. Couillard-Despres, J. M. Beaulieu and J. P. Julien (1999). Disruption of Type IV Intermediate Filament Network in Mice Lacking the Neurofilament Medium and Heavy Subunits. *J. Neurochem.* **73** (3): 972–984.

Jacot, J. G., A. D. McCulloch and J. H. Omens (2008). Substrate Stiffness Affects the Functional Maturation of Neonatal Rat Ventricular Myocytes. *Biophys. J.* **95** (7): 3479–3487.

Janmey, P. A., U. Euteneuer, P. Traub and M. Schliwa (1991). Viscoelastic Properties of Vimentin Compared with Other Filamentous Biopolymer Networks. *J. Cell Biol.* **113** (1): 155–160.

Janmey, P. A. and C. A. McCulloch (2007). Cell Mechanics: Integrating Cell Responses to Mechanical Stimuli. *Annu. Rev. Biomed. Eng.* **9**: 1–34.

Jannat, R. A., G. P. Robbins, B. G. Ricart, M. Dembo and D. A. Hammer (2010). Neutrophil Adhesion and Chemotaxis Depend on Substrate Mechanics. *J. Phys. Condensed Matter* **22** (19).

Jenkins, G. (2008). The Role of Proteases in Transforming Growth Factor-Beta Activation. *Int. J. Biochem. Cell Biol.* **40** (6–7): 1068–1078.

Jian, B., N. Narula, Q. Y. Li, E. R. Mohler, 3rd and R. J. Levy (2003). Progression of Aortic Valve Stenosis: TGF-Beta1 Is Present in Calcified Aortic Valve Cusps and Promotes Aortic Valve Interstitial Cell Calcification Via Apoptosis. *Ann. Thorac. Surg.* **75** (2): 457–465; discussion 465–456.

Johnson, C. P., H. Y. Tang, C. Carag, D. W. Speicher and D. E. Discher (2007). Forced Unfolding of Proteins within Cells. *Science* **317** (5838): 663–666.

Joshi, H. C., D. Chu, R. E. Buxbaum and S. R. Heidemann (1985). Tension and Compression in the Cytoskeleton of Pc 12 Neurites. *J. Cell Biol.* **101** (3): 697–705.

Jou, I. M., K. A. Lai, C. L. Shen and Y. Yamano (2000). Changes in Conduction, Blood Flow, Histology, and Neurological Status Following Acute Nerve-Stretch Injury Induced by Femoral Lengthening. *J. Orthop. Res.* **18** (1): 149–155.

Ju, M. S., C. C. Lin, J. L. Fan and R. J. Chen (2006). Transverse Elasticity and Blood Perfusion of Sciatic Nerves under in situ Circular Compression. *J. Biomech.* **39** (1): 97–102.

Jung, C., T. M. Chylinski, A. Pimenta, D. Ortiz and T. B. Shea (2004). Neurofilament Transport Is Dependent on Actin and Myosin. *J. Neurosci.* **24** (43): 9486–9496.

Jung, C., S. Lee, D. Ortiz, Q. Zhu, J. P. Julien and T. B. Shea (2005). The High and Middle Molecular Weight Neurofilament Subunits Regulate the Association of Neurofilaments with Kinesin: Inhibition by Phosphorylation of the High Molecular Weight Subunit. *Brain Res. Mol. Brain Res.* **141** (2): 151–155.

Kandel, E. R., J. H. Schwartz and T. M. Jessell (1991). *Principles of Neural Science*. Norwalk, Conn., Appleton & Lange.

Kaplan, N., I. K. Moore, Y. Fondufe-Mittendorf, A. J. Gossett, D. Tillo, Y. Field, et al. (2009). The DNA-Encoded Nucleosome Organization of a Eukaryotic Genome. *Nature* **458** (7236): 362–366.

Kardon, J. R. and R. D. Vale (2009). Regulators of the Cytoplasmic Dynein Motor. *Nat. Rev. Mol. Cell Biol.* **10** (12): 854–865.

Kato, M. and M. Mrksich (2004). Using Model Substrates to Study the Dependence of Focal Adhesion Formation on the Affinity of Integrin-Ligand Complexes. *Biochemistry* **43** (10): 2699–2707.

Katoh, K., Y. Kano, M. Amano, H. Onishi, K. Kaibuchi and K. Fujiwara (2001). Rho-Kinase—Mediated Contraction of Isolated Stress Fibers. *J. Cell Biol.* **153** (3): 569–584.

Kawaguchi, K. and S. Ishiwata (2000). Temperature Dependence of Force, Velocity, and Processivity of Single Kinesin Molecules. *Biochem. Biophys. Res. Commun.* **272** (3): 895–899.

Kawano, T. and S. Kidoaki (2011). Elasticity Boundary Conditions Required for Cell Mechanotaxis on Microelastically-Patterned Gels. *Biomaterials* **32** (11): 2725–2733.

Kemeny, S. F., S. Cicalese, D. S. Figueroa and A. M. Clyne (2013). Glycated Collagen and Altered Glucose Increase Endothelial Cell Adhesion Strength. *J. Cell. Physiol.* **228**: 1727–1736.

Kemeny, S. F., D. S. Figueroa, A. M. Andrews, K. A. Barbee and A. M. Clyne (2011). Glycated Collagen Alters Endothelial Cell Actin Alignment and Nitric Oxide Release in Response to Fluid Shear Stress. *J. Biomech.* **44** (10): 1927–1935.

Keski-Oja, J., K. Koli and H. von Melchner (2004). TGF-Beta Activation by Traction? *Trends Cell Biol.* **14** (12): 657–659.

Ketschek, A. and G. Gallo (2010). Nerve Growth Factor Induces Axonal Filopodia through Localized Microdomains of Phosphoinositide 3-Kinase Activity That Drive the Formation of Cytoskeletal Precursors to Filopodia. *J. Neurosci.* **30** (36): 12185–12197.

Kloxin, A. M., J. A. Benton and K. S. Anseth (2010). In situ Elasticity Modulation with Dynamic Substrates to Direct Cell Phenotype. *Biomaterials* **31** (1): 1–8.

Knight, M. M., J. van de Breevaart Bravenboer, D. A. Lee, G. J. van Osch, H. Weinans and D. L. Bader (2002). Cell and Nucleus Deformation in Compressed Chondrocyte-Alginate Constructs: Temporal Changes and Calculation of Cell Modulus. *Biochim. Biophys. Acta* **1570** (1): 1–8.

Konishi, Y. and M. Setou (2009). Tubulin Tyrosination Navigates the Kinesin-1 Motor Domain to Axons. *Nat. Neurosci.* **12** (5): 559–567.

Kramer, A., I. Liashkovich, H. Oberleithner, S. Ludwig, I. Mazur and V. Shahin (2008). Apoptosis Leads to a Degradation of Vital Components of Active Nuclear Transport and a Dissociation of the Nuclear Lamina. *Proc. Natl. Acad. Sci. U. S. A.* **105** (32): 11236–11241.

Krimm, I., C. Ostlund, B. Gilquin, J. Couprie, P. Hossenlopp, J. P. Mornon, et al. (2002). The Ig-Like Structure of the C-Terminal Domain of Lamin a/C, Mutated in Muscular Dystrophies, Cardiomyopathy, and Partial Lipodystrophy. *Structure* **10** (6): 811–823.

Krishnamurthy, V. K., F. Guilak, D. A. Narmoneva and R. B. Hinton (2011). Regional Structure-Function Relationships in Mouse Aortic Valve Tissue. *J. Biomech.* **44** (1): 77–83.

Ku, C. H., P. H. Johnson, P. Batten, P. Sarathchandra, R. C. Chambers, P. M. Taylor, et al. (2006). Collagen Synthesis by Mesenchymal Stem Cells and Aortic Valve Interstitial Cells in Response to Mechanical Stretch. *Cardiovasc. Res.* **71** (3): 548–556.

Kuhn, T., T. O. Ihalainen, J. Hyvaluoma, N. Dross, S. F. Willman, J. Langowski, et al. (2011). Protein Diffusion in Mammalian Cell Cytoplasm. *PLoS One* **6** (8): e22962.

Kull, F. J., and S. A. Endow (2013). Force Generation by Kinesin and Myosin Cytoskeletal Motor Proteins. *J. Cell. Sci.* **126**(Pt 1): 9–19.

Kunkel, T. A. and D. A. Erie (2005). DNA Mismatch Repair. *Ann. Rev. Biochem.* **74**: 681–710.

Kuo, S. C. and D. A. Lauffenburger (1993). Relationship between Receptor-Ligand Binding-Affinity and Adhesion Strength. *Biophys. J.* **65** (5): 2191–2200.

Kuo, S. C. and M. P. Sheetz (1993). Force of Single Kinesin Molecules Measured with Optical Tweezers. *Science* **260** (5105): 232–234.

Kwan, M. K., E. J. Wall, J. Massie and S. R. Garfin (1992). Strain, Stress and Stretch of Peripheral Nerve. Rabbit Experiments in vitro and in vivo. *Acta. Orthop. Scand.* **63** (3): 267–272.

Lacouture, M. E., J. L. Schaffer and L. B. Klickstein (2002). A Comparison of Type I Collagen, Fibronectin, and Vitronectin in Supporting Adhesion of Mechanically Strained Osteoblasts. *J. Bone Miner. Res.* **17** (3): 481–492.

Lammerding, J., K. N. Dahl, D. E. Discher and R. D. Kamm (2007). Nuclear Mechanics and Methods. *Methods Cell Biol.* **83**: 269–294.

Lammerding, J., L. G. Fong, J. Y. Ji, K. Reue, C. L. Stewart, S. G. Young and R. T. Lee (2006). Lamins a and C but Not Lamin B1 Regulate Nuclear Mechanics. *J. Biol. Chem.* **281** (35): 25768–25780.

Lammerding, J. and R. T. Lee (2005). The Nuclear Membrane and Mechanotransduction: Impaired Nuclear Mechanics and Mechanotransduction in Lamin a/C Deficient Cells. *Novartis Found. Symp.* **264**: 264–273; discussion 273–268.

Lammerding, J., P. C. Schulze, T. Takahashi, S. Kozlov, T. Sullivan, R. D. Kamm, et al. (2004). Lamin a/C Deficiency Causes Defective Nuclear Mechanics and Mechanotransduction. *J. Clin. Invest.* **113** (3): 370–378.

Lamoureux, P., J. Zheng, R. E. Buxbaum and S. R. Heidemann (1992). A Cytomechanical Investigation of Neurite Growth on Different Culture Surfaces. *J. Cell Biol.* **118** (3): 655–661.

Lang, F., M. Foller, K. Lang, P. Lang, M. Ritter, A. Vereninov, et al. (2007). Cell Volume Regulatory Ion Channels in Cell Proliferation and Cell Death. *Osmosensing Osmosignaling* **428**: 209–225.

LaPlaca, M. C., D. K. Cullen, J. J. McLoughlin and R. S. Cargill, 2nd (2005). High Rate Shear Strain of Three-Dimensional Neural Cell Cultures: A New in vitro Traumatic Brain Injury Model. *J. Biomech.* **38** (5): 1093–1105.

LaPlaca, M. C., G. R. Prado, D. K. Cullen and H. R. Irons (2006). High Rate Shear Insult Delivered to Cortical Neurons Produces Heterogeneous Membrane Permeability Alterations. *Conf. Proc. IEEE Eng. Med. Biol. Soc.* **1**: 2384–2387.

Layton, B. and G. Buzby (2010). An Integrated Atomic Force Microscopy Nanomanipulation Platform. USA.

Layton, B. E. (2003a). Layton's Thesis Title. *Department of Biomedical Engineering*. Ann Arbor, MI, The University of Michigan. **Ph.D.**

Layton, B. E. (2003b). Remodeling of Heterogeneous Extracellular Matrices of the Diabetic Nerve: Models and Experiments. *Mechanical Engineering and Mechanics*. Ann Arbor, MI, The University of Michigan. **PhD:** 219.

Layton, B. E., M. B. Boyd, M. S. Tripepi, B. M. Bitonti, M. N. Dollahon and R. A. Balsamo (2010). Dehydration-Induced Expression of a 31-Kda Dehydrin in Polypodium Polypodioides (Polypodiaceae) May Enable Large, Reversible Deformation of Cell Walls. *Am. J. Bot.* **97** (4): 535–544.

Layton, B. E., A. D'Souza, A. Zeiger, A. Sabur, J. Jean-Charles and W. Dampier (2008). Collagen's Triglycine Repeat Length May Help to Explain an Interdomain Transfer Event from a Eukaryote into *Trichodesmium Erythraeum*. *J. Mol. Evol.* **66** (6): 539–554.

Layton, B. E. and A. M. Sastry (2004). A Mechanical Model for Collagen Fibril Load Sharing in Peripheral Nerve of Diabetic and Nondiabetic Rats. *J. Biomech. Eng.* **126** (6): 803–814.

Layton, B. E., A. M. Sastry, C. M. Lastoskie, M. A. Philbert, T. J. Miller, K. A. Sullivan, E. L. Feldman, and W. C.-W. (2004). In situ Imaging of Mitochondrial Outer Membrane Pores Using Atomic Force Microscopy. *Biotechniques* **37**: 564–573.

Layton, B. E. and A. M. Sastry (2006). Equal and Local-Load-Sharing Micromechanical Models for Collagens: Quantitative Comparisons in Response of Non-Diabetic and Diabetic Rat Tissue. *Acta Biomater.* **2** (6): 595–607.

Leach, J. B., X. Q. Brown, J. G. Jacot, P. A. Dimilla and J. Y. Wong (2007). Neurite Outgrowth and Branching of Pc12 Cells on Very Soft Substrates Sharply Decreases Below a Threshold of Substrate Rigidity. *J. Neural. Eng.* **4** (2): 26–34.

Lebowitz, J. L. (1999). Microscopic Origins of Macroscopic Behavior. *Phys. A* **263**: 516–527.

Lee, C., J. Ma, D. N. Deal, B. P. Smith, L. A. Koman, T. L. Smith and J. S. Shilt (2006). Neuromuscular Recovery after Distraction Osteogenesis at Different Frequencies in a Rabbit Model. *J. Pediatr. Orthop.* **26** (5): 628–633.

Lee, M. K. and D. W. Cleveland (1996). Neuronal Intermediate Filaments. *Annu. Rev. Neurosci.* **19**: 187–217.

Lehenkari, P. P. and M. A. Horton (1999). Single Integrin Molecule Adhesion Forces in Intact Cells Measured by Atomic Force Microscopy. *Biochem. Biophys. Res. Commun.* **259** (3): 645–650.

Lehninger, A. L., D. L. Nelson and M. M. Cox (2005). *Lehninger Principles of Biochemistry*. New York, W. H. Freeman.

Lemaire, V., F. L. Tobin L. D. Greller, C. R. Cho and L. J. Suva (2004). Modeling the Interactions between Osteoblast and Osteoclast Activities in Bone Remodeling. *J. Theor. Biol.* **229** (3): 293–309.

Leng, M. (1985). Left-Handed Z-DNA. *Biochim. Biophys. Acta.* **825** (4): 339–344.

Lepault, J., J. L. Ranck, I. Erk and M. F. Carlier (1994). Small Angle X-Ray Scattering and Electron Cryomicroscopy Study of Actin Filaments: Role of the Bound Nucleotide in the Structure of F-Actin. *J. Struct. Biol.* **112** (1): 79–91.

Lessey, E. C., C. Guilluy and K. Burridge (2012). From Mechanical Force to Rhoa Activation. *Biochemistry* **51**(38): 7420–7432.

Leterrier, J. F. and J. Eyer (1987). Properties of Highly Viscous Gels Formed by Neurofilaments in vitro. A Possible Consequence of a Specific Inter-Filament Cross-Bridging. *Biochem. J.* **245** (1): 93–101.

Leterrier, J. F., J. Kas J. Hartwig, R. Vegners and P. A. Janmey (1996). Mechanical Effects of Neurofilament Cross-Bridges. Modulation by Phosphorylation, Lipids, and Interactions with F-Actin. *J. Biol. Chem.* **271** (26): 15687–15694.

Letourneau, P. C. (2009). Actin in Axons: Stable Scaffolds and Dynamic Filaments. *Results Probl. Cell Differ.* **48**: 65–90.

Leung, C. L., D. Sun, M. Zheng, D. R. Knowles and R. K. Liem (1999). Microtubule Actin Cross-Linking Factor (Macf): A Hybrid of Dystonin and Dystrophin That Can Interact with the Actin and Microtubule Cytoskeletons. *J. Cell Biol.* **147** (6): 1275–1286.

Levavasseur, F., Q. Zhu and J. P. Julien (1999). No Requirement of Alpha-Internexin for Nervous System Development and for Radial Growth of Axons. *Brain Res. Mol. Brain Res.* **69** (1): 104–112.

Levental, K. R., H. M. Yu, L. Kass, J. N. Lakins, M. Egeblad, J. T. Erler, et al. (2009). Matrix Crosslinking Forces Tumor Progression by Enhancing Integrin Signaling. *Cell* **139** (5): 891–906.

Levine, I. N. (2009). *Physical Chemistry*. Boston, McGraw-Hill.

Lewis, A. K. and P. C. Bridgman (1992). Nerve Growth Cone Lamellipodia Contain Two Populations of Actin Filaments That Differ in Organization and Polarity. *J. Cell Biol.* **119** (5): 1219–1243.

Li, C., Y. P. Liu, K. K. Liu and A. K. Lai (2008a). Correlations between the Experimental and Numerical Investigations on the Mechanical Properties of Erythrocyte by Laser Stretching. *IEEE Trans. Nanobiosci.* **7** (1): 80–90.

Li, F. Y., S. D. Redick, H. P. Erickson and V. T. Moy (2003). Force Measurements of the Alpha(5)Beta(1) Integrin-Fibronectin Interaction. *Biophys. J.* **84** (2): 1252–1262.

Li, J. and R. Shi (2006). A Device for the Electrophysiological Recording of Peripheral Nerves in Response to Stretch. *J. Neurosci. Methods* **154** (1–2): 102–108.

Li, P., D. Wang, J. Lucas, S. Oparil, D. Xing, X. Cao, et al. (2008b). Atrial Natriuretic Peptide Inhibits Transforming Growth Factor Beta-Induced Smad Signaling and Myofibroblast Transformation in Mouse Cardiac Fibroblasts. *Circ. Res.* **102** (2): 185–192.

Lim, R. Y. and B. Fahrenkrog (2006). The Nuclear Pore Complex up Close. *Curr. Opin. Cell Biol.* **18** (3): 342–347.

Lin, C. H. and P. Forscher (1995). Growth Cone Advance Is Inversely Proportional to Retrograde F-Actin Flow. *Neuron* **14** (4): 763–771.

Liszewski, K. (2010). New Tools Facilitate Protein Expression. *Genet. Eng. Biotechnol. N.* **30** (9): 40–41.

Liu, A. C. and A. I. Gotlieb (2008). Transforming Growth Factor-Beta Regulates in vitro Heart Valve Repair by Activated Valve Interstitial Cells. *Am. J. Pathol.* **173** (5): 1275–1285.

Liu, H., Y. Sun and C. A. Simmons (2013). Determination of Local and Global Elastic Moduli of Valve Interstitial Cells Cultured on Soft Substrates. *J. Biomech.* **46** (11): 1967–1971.

Lo, C. M., H. B. Wang, M. Dembo and Y. L. Wang (2000). Cell Movement Is Guided by the Rigidity of the Substrate. *Biophys. J.* **79** (1): 144–152.

Low, P., G. Marchand, F. Knox and P. J. Dyck (1977). Measurement of Endoneurial Fluid Pressure with Polyethylene Matrix Capsules. *Brain Res.* **122** (2): 373–377.

Lu, Y. B., T. Pannicke, E. Q. Wei, A. Bringmann, P. Wiedemann, G. Habermann, et al. (2013). Biomechanical Properties of Retinal Glial Cells: Comparative and Developmental Data. *Exp. Eye Res.* **113**: 60–65.

Luxton, G. W., E. R. Gomes, E. S. Folker, E. Vintinner and G. G. Gundersen (2010). Linear Arrays of Nuclear Envelope Proteins Harness Retrograde Actin Flow for Nuclear Movement. *Science* **329** (5994): 956–959.

Macioce, P., N. Gandolfi, C. L. Leung, S. S. Chin, F. Malchiodi-Albedi, M. Ceccarini, et al. (1999). Characterization of Nf-L and Betaiisigma1-Spectrin Interaction in Live Cells. *Exp. Cell Res.* **250** (1): 142–154.

MacKenna, D., S. R. Summerour and F. J. Villarreal (2000). Role of Mechanical Factors in Modulating Cardiac Fibroblast Function and Extracellular Matrix Synthesis. *Cardiovasc. Res.* **46** (2): 257–263.

Maekawa, S., E. Nishida, Y. Ohta and H. Sakai (1984). Isolation of Low Molecular Weight Actin-Binding Proteins from Porcine Brain. *J. Biochem.* **95** (2): 377–385.

Majhenc, J., B. Bozic, S. Svetina and B. Zeks (2004). Phospholipid Membrane Bending as Assessed by the Shape Sequence of Giant Oblate Phospholipid Vesicles. *Biochim Biophys. Acta* **1664** (2): 257–266.

Mallik, R., B. C. Carter, S. A. Lex, S. J. King and S. P. Gross (2004). Cytoplasmic Dynein Functions as a Gear in Response to Load. *Nature* **427** (6975): 649–652.

Mallik, R. and S. P. Gross (2004). Molecular Motors: Strategies to Get Along. *Curr. Biol.* **14** (22): R971–982.

Mallik, R., D. Petrov, S. A. Lex, S. J. King and S. P. Gross (2005). Building Complexity: An in vitro Study of Cytoplasmic Dynein with in vivo Implications. *Curr. Biol.* **15** (23): 2075–2085.

Malmstrom, J., H. Lindberg, C. Lindberg, C. Bratt, E. Wieslander, E. L. Delander, et al. (2004). Transforming Growth Factor-Beta 1 Specifically Induce Proteins Involved in the Myofibroblast Contractile Apparatus. *Mol. Cell Proteomics* **3** (5): 466–477.

Masszi, A. and A. Kapus (2011). Smaddening Complexity: The Role of Smad3 in Epithelial-Myofibroblast Transition. *Cells Tissues Organs* **193** (1–2): 41–52.

Masszi, A., P. Speight, E. Charbonney, M. Lodyga, H. Nakano, K. Szaszi and A. Kapus (2010). Fate-Determining Mechanisms in Epithelial-Myofibroblast Transition: Major Inhibitory Role for Smad3. *J. Cell Biol.* **188** (3): 383–399.

Matsumoto, T., H. Abe, T. Ohashi, Y. Kato and M. Sato (2002). Local Elastic Modulus of Atherosclerotic Lesions of Rabbit Thoracic Aortas Measured by Pipette Aspiration Method. *Physiol. Meas.* **23** (4): 635–648.

Mazumder, A. and G. V. Shivashankar (2007). Gold-Nanoparticle-Assisted Laser Perturbation of Chromatin Assembly Reveals Unusual Aspects of Nuclear Architecture within Living Cells. *Biophys. J.* **93** (6): 2209–2216.

McQuarrie, I. G., S. T. Brady and R. J. Lasek (1986). Diversity in the Axonal Transport of Structural Proteins: Major Differences between Optic and Spinal Axons in the Rat. *J. Neurosci.* **6** (6): 1593–1605.

Mendelson, K. and F. J. Schoen (2006). Heart Valve Tissue Engineering: Concepts, Approaches, Progress, and Challenges. *Ann. Biomed. Eng.* **34** (12): 1799–1819.

Merryman, W. D. (2007). Mechanobiology of the Aortic Valve Interstitial Cell. *School of Engineering*. Pittsburgh, University of Pittsburgh. **Doctor of Philosophy:** 196.

Merryman, W. D. (2008). Insights into (the Interstitium of) Degenerative Aortic Valve Disease. *J. Am. Coll. Cardiol.* **51** (14): 1415.

Merryman, W. D. (2010). Mechano-Potential Etiologies of Aortic Valve Disease. *J. Biomech.* **43** (1): 87–92.

Merryman, W. D., G. C. Engelmayr, J. Liao and M. S. Sacks (2006a). Defining Biomechanical Endpoints for Tissue Engineered Heart Valve Leaflets from Native Leaflet Properties. *Prog. Pediatr. Cardiol.* **21**: 153–160.

Merryman, W. D., H. Y. Huang, F. J. Schoen and M. S. Sacks (2006b). The Effects of Cellular Contraction on Aortic Valve Leaflet Flexural Stiffness. *J. Biomech.* **39** (1): 88–96.

Merryman, W. D., H. D. Lukoff, R. A. Long, G. C. Engelmayr, Jr., R. A. Hopkins and M. S. Sacks (2007). Synergistic Effects of Cyclic Tension and Transforming Growth Factor-Beta1 on the Aortic Valve Myofibroblast. *Cardiovasc. Pathol.* **16** (5): 268–276.

Merryman, W. D., I. Youn, H. D. Lukoff, P. M. Krueger, F. Guilak, R. A. Hopkins and M. S. Sacks (2006c). Correlation between Heart Valve Interstitial Cell Stiffness and Transvalvular Pressure: Implications for Collagen Biosynthesis. *Am. J. Physiol. Heart Circ. Physiol.* **290** (1): H224–H231.

Michel, D. J. and D. J. Cleaver (2007). Coarse-Grained Simulation of Amphiphilic Self-Assembly. *J. Chem. Phys.* **126** (3): 034506.

Miller, K. E. and M. P. Sheetz (2006). Direct Evidence for Coherent Low Velocity Axonal Transport of Mitochondria. *J. Cell Biol.* **173** (3): 373–381.

Millesi, H., G. Zoch and R. Reihsner (1995). Mechanical Properties of Peripheral Nerves. *Clin. Orthop. Relat. Res.* **314**: 76–83.

Mills, R. G., L. S. Minamide, A. Yuan, J. R. Bamburg and J. J. Bray (1996). Slow Axonal Transport of Soluble Actin with Actin Depolymerizing Factor, Cofilin, and Profilin Suggests Actin Moves in an Unassembled Form. *J. Neurochem.* **67** (3): 1225–1234.

Miralles, F., G. Posern, A. I. Zaromytidou and R. Treisman (2003). Actin Dynamics Control Srf Activity by Regulation of Its Coactivator Mal. *Cell* **113** (3): 329–342.

Moir, R. D., A. D. Donaldson and M. Stewart (1991). Expression in Escherichia Coli of Human Lamins a and C: Influence of Head and Tail Domains on Assembly Properties and Paracrystal Formation. *J. Cell Sci.* **99** (Pt 2): 363–372.

Moir, R. D., M. Yoon, S. Khuon and R. D. Goldman (2000). Nuclear Lamins a and B1: Different Pathways of Assembly during Nuclear Envelope Formation in Living Cells. *J. Cell Biol.* **151** (6): 1155–1168.

Montecino, M., J. L. Stein, G. S. Stein, J. B. Lian, A. J. van Wijnen, F. Cruzat, et al. (2007). Nucleosome Organization and Targeting of Swi/Snf Chromatin-Remodeling Complexes: Contributions of the DNA Sequence. *Biochem. Cell Biol.Biochimie Biol. Cell.* **85** (4): 419–425.

Monzack, E. L., X. Gu and K. S. Masters (2009). Efficacy of Simvastatin Treatment of Valvular Interstitial Cells Varies with the Extracellular Environment. *Arterioscler. Thromb. Vasc. Biol.* **29** (2): 246–253.

Mooney, D. J., R. Langer and D. E. Ingber (1995). Cytoskeletal Filament Assembly and the Control of Cell Spreading and Function by Extracellular-Matrix. *J. Cell Sci.* **108**: 2311–2320.

Mora, C., D. P. Tittensor, S. Adl, A. G. Simpson and B. Worm (2011). How Many Species Are There on Earth and in the Ocean? *PLoS Biol* **9** (8): e1001127.

Moraes, C., J. H. Chen, Y. Sun and C. A. Simmons (2010). Microfabricated Arrays for High-Throughput Screening of Cellular Response to Cyclic Substrate Deformation. *Lab Chip* **10** (2): 227–234.

Moreno-Flores, S., R. Benitez, M. Vivanco and J. L. Toca-Herrera (2010). Stress Relaxation and Creep on Living Cells with the Atomic Force Microscope: A Means to Calculate Elastic Moduli and Viscosities of Cell Components. *Nanotechnology* **21** (44): 445101.

Moreno-Herrero, F., L. Holtzer, D. A. Koster, S. Shuman, C. Dekker and N. H. Dekker (2005). Atomic Force Microscopy Shows That Vaccinia Topoisomerase Ib Generates Filaments on DNA in a Cooperative Fashion. *Nucleic Acids Res.* **33** (18): 5945–5953.

Mori, H., Y. Komiya and M. Kurokawa (1979). Slowly Migrating Axonal Polypeptides. Inequalities in Their Rate and Amount of Transport between Two Branches of Bifurcating Axons. *J. Cell Biol.* **82** (1): 174–184.

Moroz, J. D. and P. Nelson (1997). Torsional Directed Walks, Entropic Elasticity, and DNA Twist Stiffness. *Proc. Natl. Acad. Sci. U. S. A.* **94** (26): 14418–14422.

Morris, J. R. and R. J. Lasek (1984). Monomer-Polymer Equilibria in the Axon: Direct Measurement of Tubulin and Actin as Polymer and Monomer in Axoplasm. *J. Cell Biol.* **98** (6): 2064–2076.

Moshayedi, P., L. D. Costa, A. Christ, S. P. Lacour, J. Fawcett, J. Guck and K. Franze (2010). Mechanosensitivity of Astrocytes on Optimized Polyacrylamide Gels Analyzed by Quantitative Morphometry. *J. Phys.Condens. Matter* **22** (19): 194114.

Motil, J., W. K. Chan, M. Dubey, P. Chaudhury, A. Pimenta, T. M. Chylinski, et al. (2006). Dynein Mediates Retrograde Neurofilament Transport within Axons and Anterograde Delivery of Nfs from Perikarya into Axons: Regulation by Multiple Phosphorylation Events. *Cell Motil Cytoskeleton* **63** (5): 266–286.

Muller, M. J., S. Klumpp and R. Lipowsky (2008). Tug-of-War as a Cooperative Mechanism for Bidirectional Cargo Transport by Molecular Motors. *Proc. Natl. Acad. Sci. U. S. A.* **105** (12): 4609–4614.

Munevar, S., Y. L. Wang and M. Dembo (2001). Traction Force Microscopy of Migrating Normal and H-Ras Transformed 3t3 Fibroblasts. *Biophys. J.* **80** (4): 1744–1757.

Munnamalai, V. and D. M. Suter (2009). Reactive Oxygen Species Regulate F-Actin Dynamics in Neuronal Growth Cones and Neurite Outgrowth. *J. Neurochem.* **108** (3): 644–661.

Myers, K. A. and P. W. Baas (2007). Kinesin-5 Regulates the Growth of the Axon by Acting as a Brake on Its Microtubule Array. *J. Cell Biol.* **178** (6): 1081–1091.

Myers, K. A., I. Tint, C. V. Nadar, Y. He, M. M. Black and P. W. Baas (2006). Antagonistic Forces Generated by Cytoplasmic Dynein and Myosin-Ii during Growth Cone Turning and Axonal Retraction. *Traffic* **7** (10): 1333–1351.

Myllyharju, J. and K. I. Kivirikko (2001). Collagens and Collagen-Related Diseases. *Ann. Med.* **33** (1): 7–21.

Nadar, V. C., A. Ketschek, K. A. Myers, G. Gallo and P. W. Baas (2008). Kinesin-5 Is Essential for Growth-Cone Turning. *Curr. Biol.* **18** (24): 1972–1977.

Nadkarni, S. K., B. E. Bouma, T. Helg, R. Chan, E. Halpern, A. Chau, et al. (2005). Characterization of Atherosclerotic Plaques by Laser Speckle Imaging. *Circulation* **112** (6): 885–892.

Nahon, P., A. Kettaneh, M. Lemoine, O. Seror, N. Barget, J. C. Trinchet, et al. (2009). Liver Stiffness Measurement in Patients with Cirrhosis and Hepatocellular Carcinoma: A Case-Control Study. *Eur. J. Gastroenterol. Hepatol.* **21** (2): 214–219.

Nakata, T. and H. Yorifuji (1999). Morphological Evidence of the Inhibitory Effect of Taxol on the Fast Axonal Transport. *Neurosci. Res.* **35** (2): 113–122.

Narayanareddy, B. R., S. Vartiainen, N. Hariri, D. K. O'Dowd, and S. P. Gross (2014). A Biophysical Analysis of Mitochondrial Movement: Differences between Transport in Neuronal Cell Bodies versus Processes. *Traffic* **15**(7): 762–771.

Nee, S. (2003). Unveiling Prokaryotic Diversity. *Trends Ecol. Evol.* **18** (2): 62–63.

Nemir, S., H. N. Hayenga and J. L. West (2010). Pegda Hydrogels with Patterned Elasticity: Novel Tools for the Study of Cell Response to Substrate Rigidity. *Biotechnol. Bioeng.* **105** (3): 636–644.

Netti, P. A., D. A. Berk, M. A. Swartz, A. J. Grodzinsky and R. K. Jain (2000). Role of Extracellular Matrix Assembly in Interstitial Transport in Solid Tumors. *Cancer Res.* **60** (9): 2497–2503.

Newport, J. W., K. L. Wilson and W. G. Dunphy (1990). A Lamin-Independent Pathway for Nuclear Envelope Assembly. *J. Cell Biol.* **111** (6 Pt 1): 2247–2259.

Niklas, K. J. (2014). The Evolutionary-Developmental Origins of Multicellularity. *Am. J. Bot.* **101**(1): 6–25.

Nikolova, V., C. Leimena, A. C. McMahon, J. C. Tan, S. Chandar, D. Jogia, et al. (2004). Defects in Nuclear Structure and Function Promote Dilated Cardiomyopathy in Lamin a/C-Deficient Mice. *J. Clin. Invest.* **113** (3): 357–369.

O'Brien, K. D., D. D. Reichenbach, S. M. Marcovina, J. Kuusisto, C. E. Alpers and C. M. Otto (1996). Apolipoproteins B, (a), and E Accumulate in the Morphologically Early Lesion of 'Degenerative' Valvular Aortic Stenosis. *Arterioscler. Thromb. Vasc. Biol.* **16** (4): 523–532.

O'Toole, M., P. Lamoureux and K. E. Miller (2008). A Physical Model of Axonal Elongation: Force, Viscosity, and Adhesions Govern the Mode of Outgrowth. *Biophys. J.* **94** (7): 2610–2620.

Oakes, P. W., D. C. Patel, N. A. Morin, D. P. Zitterbart, B. Fabry, J. S. Reichner and J. X. Tang (2009). Neutrophil Morphology and Migration Are Affected by Substrate Elasticity. *Blood* **114** (7): 1387–1395.

Ohashi, T., M. Hagiwara, D. L. Bader and M. M. Knight (2006). Intracellular Mechanics and Mechanotransduction Associated with Chondrocyte Deformation during Pipette Aspiration. *Biorheology* **43** (3–4): 201–214.

Ohayon, J., G. Finet, A. M. Gharib, D. A. Herzka, P. Tracqui, J. Heroux, et al. (2008). Necrotic Core Thickness and Positive Arterial Remodeling Index: Emergent Biomechanical Factors for Evaluating the Risk of Plaque Rupture. *Am. J. Physiol. Heart Circ. Physiol.* **295** (2): H717–H727.

Okabe, S. and N. Hirokawa (1990). Turnover of Fluorescently Labelled Tubulin and Actin in the Axon. *Nature* **343** (6257): 479–482.

Okabe, S. and N. Hirokawa (1992). Differential Behavior of Photoactivated Microtubules in Growing Axons of Mouse and Frog Neurons. *J. Cell Biol.* **117** (1): 105–120.

Ophir, J., I. Cespedes, H. Ponnekanti, Y. Yazdi and X. Li (1991). Elastography—a Quantitative Method for Imaging the Elasticity of Biological Tissues. *Ultrason. Imaging* **13** (2): 111–134.

Oren, A. (2004). Prokaryote Diversity and Taxonomy: Current Status and Future Challenges. *Philos. Trans. R. Soc. Lond. Series B Biol. Sci.* **359** (1444): 623–638.

Orr, A. W., M. H. Ginsberg, S. J. Shattil, H. Deckmyn and M. A. Schwartz (2006). Matrix-Specific Suppression of Integrin Activation in Shear Stress Signaling. *Mol. Biol. Cell* **17** (11): 4686–4697.

Ostashevsky, J. (2002). A Polymer Model for Large-Scale Chromatin Organization in Lower Eukaryotes. *Mol. Biol. Cell* **13** (6): 2157–2169.

Oster (1970). Some Applications of Detailed Balancing. *Am. J. Phys.* **38** (6): 754.

Otto, C. M., J. Kuusisto, D. D. Reichenbach, A. M. Gown and K. D. O'Brien (1994). Characterization of the Early Lesion of "Degenerative" Valvular Aortic Stenosis. Histological and Immunohistochemical Studies. *Circulation* **90** (2): 844–853.

Otto, C. M., B. K. Lind, D. W. Kitzman, B. J. Gersh and D. S. Siscovick (1999). Association of Aortic-Valve Sclerosis with Cardiovascular Mortality and Morbidity in the Elderly. *N. Engl. J. Med.* **341** (3): 142–147.

Ouyang, H., B. Galle, J. Li, E. Nauman and R. Shi (2008). Biomechanics of Spinal Cord Injury: A Multimodal Investigation Using *ex vivo* Guinea Pig Spinal Cord White Matter. *J. Neurotrauma* **25** (1): 19–29.

Pajerowski, J. D., K. N. Dahl, F. L. Zhong, P. J. Sammak and D. E. Discher (2007). Physical Plasticity of the Nucleus in Stem Cell Differentiation. *Proc. Natl. Acad. Sci. U. S. A.* **104** (40): 15619–15624.

Palecek, S. P., J. C. Loftus, M. H. Ginsberg, D. A. Lauffenburger and A. F. Horwitz (1997). Integrin-Ligand Binding Properties Govern Cell Migration Speed through Cell-Substratum Adhesiveness. *Nature* **385** (6616): 537–540.

Parada, L. A., S. Sotiriou and T. Misteli (2004). Spatial Genome Organization. *Exp. Cell Res.* **296** (1): 64–70.

Park, S. and Y. J. Lee (2013). Nano-Mechanical Compliance of Muller Cells Investigated by Atomic Force Microscopy. *Int. J. Biol. Sci.* **9** (7): 702–706.

Park, P. J., and K. J. Lee (2013). A Modified Active Brownian Dynamics Model Using Asymmetric Energy Conversion and Its Application to the Molecular Motor System. *J. Biol. Phys.* **39**(3): 439–452.

Paszek, M. J. and V. M. Weaver (2004). The Tension Mounts: Mechanics Meets Morphogenesis and Malignancy. *J. Mammary Gland Biol. Neoplasia* **9** (4): 325–342.

Paszek, M. J., N. Zahir, K. R. Johnson, J. N. Lakins, G. I. Rozenberg, A. Gefen, et al. (2005). Tensional Homeostasis and the Malignant Phenotype. *Cancer Cell* **8** (3): 241–254.

Pederson, T. and U. Aebi (2002). Actin in the Nucleus: What Form and What For? *J. Struct. Biol.* **140** (1–3): 3–9.

Pelham, R. J. and Y. L. Wang (1997). Cell Locomotion and Focal Adhesions Are Regulated by Substrate Flexibility. *Proc. Natl. Acad. Sci. U. S. A.* **94** (25): 13661–13665.

Peltonen, T. O., P. Taskinen, Y. Soini, J. Rysa, J. Ronkainen, P. Ohtonen, et al. (2007). Distinct Downregulation of C-Type Natriuretic Peptide System in Human Aortic Valve Stenosis. *Circulation* **116** (11): 1283–1289.

Pesen, D. and J. H. Hoh (2005). Micromechanical Architecture of the Endothelial Cell Cortex. *Biophys. J.* **88** (1): 670–679.

Peyton, S. R. and A. J. Putnam (2005). Extracellular Matrix Rigidity Governs Smooth Muscle Cell Motility in a Biphasic Fashion. *J. Cell. Physiol.* **204** (1): 198–209.

Peyton, S. R., C. B. Raub, V. P. Keschrumrus and A. J. Putnam (2006). The Use of Poly(Ethylene Glycol) Hydrogels to Investigate the Impact of Ecm Chemistry and Mechanics on Smooth Muscle Cells. *Biomaterials* **27** (28): 4881–4893.

Pfister, B. J., J. H. Huang, N. Kameswaran, E. L. Zager and D. H. Smith (2007). Neural Engineering to Produce in vitro Nerve Constructs and Neurointerface. *Neurosurgery* **60** (1): 137–141; discussion 141–132.

Pfister, B. J., A. Iwata, D. F. Meaney and D. H. Smith (2004). Extreme Stretch Growth of Integrated Axons. *J. Neurosci.* **24** (36): 7978–7983.

Pfister, B. J., T. P. Weihs, M. Betenbaugh and G. Bao (2003). An *in vitro* Uniaxial Stretch Model for Axonal Injury. *Ann. Biomed. Eng.* **31** (5): 589–598.

Pfister, K. K. (1999). Cytoplasmic Dynein and Microtubule Transport in the Axon: The Action Connection. *Mol. Neurobiol.* **20** (2–3): 81–91.

Pho, M., W. Lee, D. R. Watt, C. Laschinger, C. A. Simmons and C. A. McCulloch (2008). Cofilin Is a Marker of Myofibroblast Differentiation in Cells from Porcine Aortic Cardiac Valves. *Am. J. Physiol.Heart Circ. Physiol.* **294** (4): H1767–1778.

Polte, T. R., G. S. Eichler, N. Wang and D. E. Ingber (2004). Extracellular Matrix Controls Myosin Light Chain Phosphorylation and Cell Contractility through Modulation of Cell Shape and Cytoskeletal Prestress. *Am. J. Physiol. Cell Physiol.* **286** (3): C518–C528.

Polyakova, V., S. Hein, S. Kostin, T. Ziegelhoeffer and J. Schaper (2004). Matrix Metalloproteinases and Their Tissue Inhibitors in Pressure-Overloaded Human Myocardium during Heart Failure Progression. *J. Am. Coll. Cardiol.* **44** (8): 1609–1618.

Popov, S., A. Brown and M. M. Poo (1993). Forward Plasma Membrane Flow in Growing Nerve Processes. *Science* **259** (5092): 244–246.

Previtera, M. L., C. G. Langhammer and B. L. Firestein (2010a). Effects of Substrate Stiffness and Cell Density on Primary Hippocampal Cultures. *J. Biosci. Bioeng.* **110** (4): 459–470.

Previtera, M. L., C. G. Langhammer, N. A. Langrana and B. L. Firestein (2010b). Regulation of Dendrite Arborization by Substrate Stiffness Is Mediated by Glutamate Receptors. *Ann. Biomed. Eng.* **38** (12): 3733–3743.

Price, S. (2013). Investigating the Role of the Collagen Protein in *Trichodesmium Erythraeum. Department of Biology.* Philadelphia, Drexel University. **PhD.**

Price, S. and S. Anandan (2013). Characterization of a Novel Collagen-Like Protein Trpa in the Cyanobacterium Trichodesmium Erythraeum Ims 101. *J. Phycol.* **49**: 758–764.

Pullarkat, P. A., P. Dommersnes, P. Fernandez, J. F. Joanny and A. Ott (2006). Osmotically Driven Shape Transformations in Axons. *Phys. Rev. Lett.* **96** (4): 048104.

Purcell, E. M. (1977). Life at Low Reynolds Number. *Am. J. Phys.* **45** (1): 1–9.

Qin, Z. and M. J. Buehler (2011). Flaw Tolerance of Nuclear Intermediate Filament Lamina under Extreme Mechanical Deformation. *ACS Nano* **5** (4): 3034–3042.

Qin, Z., M. J. Buehler and L. Kreplak (2010). A Multi-Scale Approach to Understand the Mechanobiology of Intermediate Filaments. *J. Biomech.* **43** (1): 15–22.

Qin, Z., L. Kreplak and M. J. Buehler (2009). Hierarchical Structure Controls Nanomechanical Properties of Vimentin Intermediate Filaments. *PLoS One* **4** (10): e7294.

Quist, A. P., S. K. Rhee, H. Lin and R. Lal (2000). Physiological Role of Gap-Junctional Hemichannels: Extracellular Calcium-Dependent Isosmotic Volume Regulation. *J. Cell Biol.* **148** (5): 1063–1074.

Rabkin-Aikawa, E., M. Farber, M. Aikawa and F. J. Schoen (2004). Dynamic and Reversible Changes of Interstitial Cell Phenotype during Remodeling of Cardiac Valves. *J. Heart Valve Dis.* **13** (5): 841–847.

Radmacher, M. (2002). Measuring the Elastic Properties of Living Cells by the Atomic Force Microscope. *Methods Cell Biol.* **68**: 67–90.

Raharjo, W. H., P. Enarson, T. Sullivan, C. L. Stewart and B. Burke (2001). Nuclear Envelope Defects Associated with Lmna Mutations Cause Dilated Cardiomyopathy and Emery-Dreifuss Muscular Dystrophy. *J. Cell Sci.* **114** (Pt 24): 4447–4457.

Rahimi, M. and M. Arroyo (2012). Shape Dynamics, Lipid Hydrodynamics, and the Complex Viscoelasticity of Bilayer Membranes [Corrected]. *Phys. Rev. E Stat. Nonlin. Soft Matter Phys.* **86** (1 Pt 1): 011932.

Rajamannan, N. M., F. J. Evans, E. Aikawa, K. J. Grande-Allen, L. L. Demer, D. D. Heistad, et al. (2011). Calcific Aortic Valve Disease: Not Simply a Degenerative Process: A Review and Agenda for Research from the National Heart and Lung and Blood Institute Aortic Stenosis Working Group—Executive Summary: Calcific Aortic Valve Disease-2011 Update. *Circulation* **124** (16): 1783–1791.

Ramachandran, S., A. P. Quist, S. Kumar and R. Lal (2006). Cisplatin Nanoliposomes for Cancer Therapy: Afm and Fluorescence Imaging of Cisplatin Encapsulation, Stability, Cellular Uptake, and Toxicity. *Langmuir* **22** (19): 8156–8162.

Rammensee, S., P. A. Janmey and A. R. Bausch (2007). Mechanical and Structural Properties of in vitro Neurofilament Hydrogels. *Eur. Biophys. J.* **36** (6): 661–668.

Rao, M. V., J. Campbell, A. Yuan, A. Kumar, T. Gotow, Y. Uchiyama and R. A. Nixon (2003). The Neurofilament Middle Molecular Mass Subunit Carboxyl-Terminal Tail Domains Is Essential for the Radial Growth and Cytoskeletal Architecture of Axons but Not for Regulating Neurofilament Transport Rate. *J. Cell Biol.* **163** (5): 1021–1031.

Rao, M. V., L. J. Engle, P. S. Mohan, A. Yuan, D. Qiu, A. Cataldo, et al. (2002a). Myosin Va Binding to Neurofilaments Is Essential for Correct Myosin Va Distribution and Transport and Neurofilament Density. *J. Cell Biol.* **159** (2): 279–290.

Rao, M. V., M. L. Garcia, Y. Miyazaki, T. Gotow, A. Yuan, S. Mattina, et al. (2002b). Gene Replacement in Mice Reveals That the Heavily Phosphorylated Tail of Neurofilament Heavy Subunit Does Not Affect Axonal Caliber or the Transit of Cargoes in Slow Axonal Transport. *J. Cell Biol.* **158** (4): 681–693.

Reinhart-King, C. A. (2011). How Matrix Properties Control the Self-Assembly and Maintenance of Tissues. *Ann. Biomed. Eng.* **39** (7): 1849–1856.

Ren, J., S. Yu, N. Gao and Q. Zou (2013). Indentation Quantification for in-Liquid Nanomechanical Measurement of Soft Material Using an Atomic Force Microscope: Rate-Dependent Elastic Modulus of Live Cells. *Phys. Rev. E Stat. Nonlin. Soft Matter Phys.* **88** (5): 052711.

Rhee, S. K., A. P. Quist and R. Lal (1998). Amyloid Beta Protein-(1–42) Forms Calcium-Permeable, Zn^{2+}-Sensitive Channel. *J. Biol. Chem.* **273** (22): 13379–13382.

Ribeiro, A., S. Tottey, R. Taylor, R. Bise, T. Kanade, S. Badylak and D. KN (2012). Mechanical Characterization of Adult Stem Cells from Bone Marrow and Perivascular Niches. *J. Biomech.* **45** (7): 1280–1287.

Ricard-Blum, S. (2011). The Collagen Family. *Cold Spring Harb. Perspect. Biol.* **3**: a004978.

Richmond, T. J., J. T. Finch, B. Rushton, D. Rhodes and A. Klug (1984). Structure of the Nucleosome Core Particle at 7 a Resolution. *Nature* **311** (5986): 532–537.

Ridley, A. J. and A. Hall (1992). The Small Gtp-Binding Protein Rho Regulates the Assembly of Focal Adhesions and Actin Stress Fibers in Response to Growth Factors. *Cell* **70** (3): 389–399.

Rodriguez, K. J. and K. S. Masters (2009). Regulation of Valvular Interstitial Cell Calcification by Components of the Extracellular Matrix. *J. Biomed. Mater. Res. A* **90** (4): 1043–1053.

Roger, V. L., A. S. Go, D. M. Lloyd-Jones, R. J. Adams, J. D. Berry, T. M. Brown, et al. (2011). Heart Disease and Stroke Statistics—2011 Update: A Report from the American Heart Association. *Circulation* **123** (4): e18-e209.

Roingeard, P. (2008). Viral Detection by Electron Microscopy: Past, Present and Future. *Biol. Cell* **100** (8): 491–501.

Rosenhek, R., F. Rader, N. Loho, H. Gabriel, M. Heger, U. Klaar, et al. (2004). Statins but Not Angiotensin-Converting Enzyme Inhibitors Delay Progression of Aortic Stenosis. *Circulation* **110** (10): 1291–1295.

Rossebo, A. B., T. R. Pedersen, K. Boman, P. Brudi, J. B. Chambers, K. Egstrup, et al. (2008). Intensive Lipid Lowering with Simvastatin and Ezetimibe in Aortic Stenosis. *N. Engl. J. Med.* **359** (13): 1343–1356.

Rossmann, M. G., V. V. Mesyanzhinov, F. Arisaka and P. G. Leiman (2004). The Bacteriophage T4 DNA Injection Machine. *Curr. Opin. Struct. Biol.* **14** (2): 171–180.

Rothenberg, M. E., S. L. Rogers, R. D. Vale, L. Y. Jan and Y. N. Jan (2003). Drosophila Pod-1 Crosslinks Both Actin and Microtubules and Controls the Targeting of Axons. *Neuron* **39** (5): 779–791.

Rotsch, C. and M. Radmacher (2000). Drug-Induced Changes of Cytoskeletal Structure and Mechanics in Fibroblasts: An Atomic Force Microscopy Study. *Biophys. J.* **78** (1): 520–535.

Rowat, A. (2009). Physical Properties of the Nucleus Studied by Micropipette Aspiration. *Methods Mol. Biol.* **464**: 3–12.

Rowat, A., L. Foster, M. Nielsen, M. Weiss and J. Ipsen (2005). Characterization of the Elastic Properties of the Nuclear Envelope. *J. R. Soc. Interface* **2** (2): 63–69.

Rouch, D. A. (2014). Evolution of the First Genetic Cells and the Universal Genetic Code: A Hypothesis Based on Macromolecular Coevolution of RNA and Proteins. *J. Theor. Biol.* **357**: 220–244.

Ruozi, B., G. Tosi, F. Forni, M. Fresta and M. A. Vandelli (2005). Atomic Force Microscopy and Photon Correlation Spectroscopy: Two Techniques for Rapid Characterization of Liposomes. *Eur. J. Pharm. Sci.* **25**: 81–89.

Ruozi, B., G. Tosi, E. Leo and M. A. Vandelli (2007). Application of Atomic Force Microscopy to Characterize Liposomes as Drug and Gene Carriers. *Talanta* **73** (12–22).

Ruthel, G. and G. Banker (1998). Actin-Dependent Anterograde Movement of Growth-Cone-Like Structures Along Growing Hippocampal Axons: A Novel Form of Axonal Transport? *Cell Motil Cytoskeleton* **40** (2): 160–173.

Ruthel, G. and G. Banker (1999). Role of Moving Growth Cone-Like "Wave" Structures in the Outgrowth of Cultured Hippocampal Axons and Dendrites. *J. Neurobiol.* **39** (1): 97–106.

Rydevik, B. L., M. K. Kwan, R. R. Myers, R. A. Brown, K. J. Triggs, S. L. Woo and S. R. Garfin (1990). An in vitro Mechanical and Histological Study of Acute Stretching on Rabbit Tibial Nerve. *J. Orthop. Res.* **8** (5): 694–701.

Sacks, M. S., W. David Merryman and D. E. Schmidt (2009). On the Biomechanics of Heart Valve Function. *J. Biomech.* **42** (12): 1804–1824.

Sanchez-Soriano, N., M. Travis, F. Dajas-Bailador, C. Goncalves-Pimentel, A. J. Whitmarsh and A. Prokop (2009). Mouse Acf7 and Drosophila Short Stop Modulate Filopodia Formation and Microtubule Organisation during Neuronal Growth. *J. Cell Sci.* **122** (Pt 14): 2534–2542.

Sandh, G., L. Xu and B. Bergman (2012). Diazocyte Development in the Marine Diazotrophic Cyanobacterium Trichodesmium. *Microbiology* **158** (Pt 2): 345–352.

Sandrin, L., B. Fourquet, J. M. Hasquenoph, S. Yon, C. Fournier, F. Mal, et al. (2003). Transient Elastography: A New Noninvasive Method for Assessment of Hepatic Fibrosis. *Ultrasound Med. Biol.* **29** (12): 1705–1713.

Satta, J., J. Oiva, T. Salo, H. Eriksen, P. Ohtonen, F. Biancari, et al. (2003). Evidence for an Altered Balance between Matrix Metalloproteinase-9 and Its Inhibitors in Calcific Aortic Stenosis. *Ann. Thorac. Surg.* **76** (3): 681–688; discussion 688.

Scaffidi, P. and T. Misteli (2005). Reversal of the Cellular Phenotype in the Premature Aging Disease Hutchinson-Gilford Progeria Syndrome. *Nat. Med.* **11** (4): 440–445.

Schaefer, A. W., N. Kabir and P. Forscher (2002). Filopodia and Actin Arcs Guide the Assembly and Transport of Two Populations of Microtubules with Unique Dynamic Parameters in Neuronal Growth Cones. *J. Cell Biol.* **158** (1): 139–152.

Schirmer, E. C. and L. Gerace (2005). The Nuclear Membrane Proteome: Extending the Envelope. *Trends Biochem. Sci.* **30** (10): 551–558.

Schnabl, B., Y. O. Kweon, J. P. Frederick, X. F. Wang, R. A. Rippe and D. A. Brenner (2001). The Role of Smad3 in Mediating Mouse Hepatic Stellate Cell Activation. *Hepatology* **34** (1): 89–100.

Schoen, F. J. (1997). Aortic Valve Structure-Function Correlations: Role of Elastic Fibers No Longer a Stretch of the Imagination. *J. Heart Valve Dis.* **6** (1): 1–6.

Schoen, F. J. (2008). Evolving Concepts of Cardiac Valve Dynamics: The Continuum of Development, Functional Structure, Pathobiology, and Tissue Engineering. *Circulation* **118** (18): 1864–1880.

Schoen, F. J. (2011). Mechanisms of Function and Disease of Natural and Replacement Heart Valves. *Annu. Rev. Pathol.* **7**: 161–183.

Schrödinger, E. (2012). *What Is Life?: The Physical Aspect of the Living Cell: With, Mind and Matter and Autobiographical Sketches.* Cambridge; New York, Cambridge University Press.

Schurch, W., T. A. Seemayer, B. Hinz and G. Gabbiani (2007). Myofibroblast. *Histology for Pathologists.* M. SE. Philadelphia, USA, Lippincott-Williams & Wilkins: 123–164.

Scott, K. A., P. J. Bond, A. Ivetac, A. P. Chetwynd, S. Khalid and M. S. P. Sansom (2008). Coarse-Grained Md Simulations of Membrane Protein-Bilayer Self-Assembly. *Structure* **16** (4): 621–630.

Segal, E., Y. Fondufe-Mittendorf, L. Chen, A. Thastrom, Y. Field, I. K. Moore, et al. (2006). A Genomic Code for Nucleosome Positioning. *Nature* **442** (7104): 772–778.

Segura-Totten, M., A. K. Kowalski, R. Craigie and K. L. Wilson (2002). Barrier-to-Autointegration Factor: Major Roles in Chromatin Decondensation and Nuclear Assembly. *J. Cell Biol.* **158** (3): 475–485.

Semler, E. J., P. A. Lancin, A. Dasgupta and P. V. Moghe (2005). Engineering Hepatocellular Morphogenesis and Function Via Ligand-Presenting Hydrogels with Graded Mechanical Compliance. *Biotechnol. Bioeng.* **89** (3): 296–307.

Senior, A. and L. Gerace (1988). Integral Membrane Proteins Specific to the Inner Nuclear Membrane and Associated with the Nuclear Lamina. *J. Cell Biol.* **107** (6 Pt 1): 2029–2036.

Sept, D. and F. C. MacKintosh (2010). Microtubule Elasticity: Connecting All-Atom Simulations with Continuum Mechanics. *Phys. Rev. Lett.* **104**: 018101.

Sewell-Loftin, M. K. C. B. Brown, H. S. Baldwin and W. D. Merryman (2012). A Novel Technique for Quantifying Mouse Heart Valve

Leaflet Stiffness with Atomic Force Microscopy. *J. Heart Valve Dis.* **21** (4): 513–520.

Shahin, V. (2006). Route of Glucocorticoid-Induced Macromolecules across the Nuclear Envelope as Viewed by Atomic Force Microscopy. *Pflugers Arch.* **453** (1): 1–9.

Shavelle, D. M., J. Takasu, M. J. Budoff, S. Mao, X. Q. Zhao and K. D. O'Brien (2002). Hmg Coa Reductase Inhibitor (Statin) and Aortic Valve Calcium. *Lancet* **359** (9312): 1125–1126.

Shea, T. B. (2000). Microtubule Motors, Phosphorylation and Axonal Transport of Neurofilaments. *J. Neurocytol.* **29** (11–12): 873–887.

Sheppard, D. (2005). Integrin-Mediated Activation of Latent Transforming Growth Factor Beta. *Cancer Metastasis Rev.* **24** (3): 395–402.

Sheetz, M. P., and J. A. Spudich (1983). Movement of Myosin-Coated Fluorescent Beads on Actin Cables in vitro. *Nature* **303**(5912): 31–35.

Shieh, A. C. (2011). Biomechanical Forces Shape the Tumor Microenvironment. *Ann. Biomed. Eng.* **39** (5): 1379–1389.

Shimizu, H., R. Aufschnaiter, L. Li, M. P. Sarras, D. B. Borza, D. R. Abrahamson, et al. (2008). The Extracellular Matrix of Hydra Is a Porous Sheet and Contains Type IV Collagen. *Zoology* **111** (5): 410–418.

Shubin, N. (2013). *The Universe Within: Discovering the Common History of Rocks, Planets, and People*. New York, Pantheon Books.

Sider, K. L., M. C. Blaser and C. A. Simmons (2011). Animal Models of Calcific Aortic Valve Disease. *Int. J. Inflam.* **2011**: 364310.

Simmons, C. A., G. R. Grant, E. Manduchi and P. F. Davies (2005). Spatial Heterogeneity of Endothelial Phenotypes Correlates with Side-Specific Vulnerability to Calcification in Normal Porcine Aortic Valves. *Circ. Res.* **96** (7): 792–799.

Sirghi, L., J. Ponti, F. Broggi and F. Rossi (2008). Probing Elasticity and Adhesion of Live Cells by Atomic Force Microscopy Indentation. *Eur Biophys J.* **37** (6): 935–945.

Slaughter, T., J. Wang and M. M. Black (1997). Microtubule Transport from the Cell Body into the Axons of Growing Neurons. *J. Neurosci.* **17** (15): 5807–5819.

Smith, D. H., J. A. Wolf and D. F. Meaney (2001). A New Strategy to Produce Sustained Growth of Central Nervous System Axons: Continuous Mechanical Tension. *Tissue Eng.* **7** (2): 131–139.

Smith, S. B., Y. Cui and C. Bustamante (1996). Overstretching B-DNA: The Elastic Response of Individual Double-Stranded and Single-Stranded DNA Molecules. *Science* **271** (5250): 795–799.

Smith, S. B., L. Finzi and C. Bustamante (1992). Direct Mechanical Measurements of the Elasticity of Single DNA Molecules by Using Magnetic Beads. *Science* **258** (5085): 1122–1126.

Solon, J., I. Levental, K. Sengupta, P. C. Georges and P. A. Janmey (2007). Fibroblast Adaptation and Stiffness Matching to Soft Elastic Substrates. *Biophys. J.* **93** (12): 4453–4461.

Spanakis, S. G., S. Petridou and S. K. Masur (1998). Functional Gap Junctions in Corneal Fibroblasts and Myofibroblasts. *Invest. Ophthalmol. Vis. Sci.* **39** (8): 1320–1328.

Spiegel, D. A., A. V. Seaber, L. E. Chen and J. R. Urbaniak (1993). Recovery Following Stretch Injury to the Sciatic Nerve of the Rat: An in vivo Study. *J. Reconstr. Microsurg.* **9** (1): 69–74.

Srigunapalan, S., C. Lam, A. R. Wheeler and C. A. Simmons (2011). A Microfluidic Membrane Device to Mimic Critical Components of the Vascular Microenvironment. *Biomicrofluidics* **5** (1): 13409.

Stewart, B. F., D. Siscovick, B. K. Lind, J. M. Gardin, J. S. Gottdiener, V. E. Smith, et al. (1997). Clinical Factors Associated with Calcific Aortic Valve Disease. Cardiovascular Health Study. *J. Am. Coll. Cardiol.* **29** (3): 630–634.

Stewart, C. L., S. Kozlov, L. G. Fong and S. G. Young (2007). Mouse Models of the Laminopathies. *Exp. Cell Res.* **313** (10): 2144–2156.

Stierle, V., J. Couprie, C. Ostlund, I. Krimm, S. Zinn-Justin, P. Hossenlopp, et al. (2003). The Carboxyl-Terminal Region Common to Lamins a and C Contains a DNA Binding Domain. *Biochemistry* **42** (17): 4819–4828.

Stiess, M., N. Maghelli, L. C. Kapitein, S. Gomis-Ruth, M. Wilsch-Brauninger, C. C. Hoogenraad, et al. (2010). Axon Extension Occurs Independently of Centrosomal Microtubule Nucleation. *Science* **327** (5966): 704–707.

Storm, C. and P. C. Nelson (2003). Theory of High-Force DNA Stretching and Overstretching. *Physical Review E: Statistical, Nonlinear, and Soft Matter Physics* **67** (5 Pt 1): 051906.

Strelkov, S. V., J. Schumacher, P. Burkhard, U. Aebi and H. Herrmann (2004). Crystal Structure of the Human Lamin a Coil 2b Dimer: Implications for the Head-to-Tail Association of Nuclear Lamins. *J. Mol. Biol.* **343** (4): 1067–1080.

Stuurman, N., S. Heins and U. Aebi (1998). Nuclear Lamins: Their Structure, Assembly, and Interactions. *J. Struct. Biol.* **122** (1–2): 42–66.

Sullivan, T., D. Escalante-Alcalde, H. Bhatt, M. Anver, N. Bhat, K. Nagashima, et al. (1999). Loss of a-Type Lamin Expression Compromises Nuclear Envelope Integrity Leading to Muscular Dystrophy. *J. Cell Biol.* **147** (5): 913–920.

Sun, H. Q., K. Kwiatkowska and H. L. Yin (1995). Actin Monomer Binding Proteins. *Curr. Opin. Cell Biol.* **7** (1): 102–110.

Svetina, S. and B. Zeks (2002). Shape Behavior of Lipid Vesicles as the Basis of Some Cellular Processes. *Anat. Rec.* **268** (3): 215–225.

Svoboda, K., C. F. Schmidt, B. J. Schnapp, and S. M. Block (1993). Direct Observation of Kinesin Stepping by Optical Trapping Interferometry. *Nature* **365**(6448): 721–727.

Tang-Schomer, M. D., V. E. Johnson, P. W. Baas, W. Stewart and D. H. Smith (2012). Partial Interruption of Axonal Transport Due to Microtubule Breakage Accounts for the Formation of Periodic Varicosities after Traumatic Axonal Injury. *Exp. Neurol.* **233** (1): 364–372.

Tashiro, T. and Y. Komiya (1989). Stable and Dynamic Forms of Cytoskeletal Proteins in Slow Axonal Transport. *J. Neurosci.* **9** (3): 760–768.

Tashiro, T. and Y. Komiya (1991). Changes in Organization and Axonal Transport of Cytoskeletal Proteins during Regeneration. *J. Neurochem.* **56** (5): 1557–1563.

Taubenberger, A., D. A. Cisneros, J. Friedrichs, P. H. Puech, D. J. Muller and C. M. Franz (2007). Revealing Early Steps of Alpha(2)Beta(1) Integrin-Mediated Adhesion to Collagen Type I by Using Single-Cell Force Spectroscopy. *Mol. Biol. Cell* **18** (5): 1634–1644.

Taylor, P. M., S. P. Allen and M. H. Yacoub (2000). Phenotypic and Functional Characterization of Interstitial Cells from Human Heart Valves, Pericardium and Skin. *J. Heart Valve Dis.* **9** (1): 150–158.

Taylor, P. M., P. Batten, N. J. Brand, P. S. Thomas and M. H. Yacoub (2003). The Cardiac Valve Interstitial Cell. *Int. J. Biochem. Cell Biol.* **35** (2): 113–118.

ten Dijke, P. and H. M. Arthur (2007). Extracellular Control of TGF-Beta Signalling in Vascular Development and Disease. *Nat. Rev. Mol. Cell Biol.* **8** (11): 857–869.

Toepfer, C., and J. Sellers (2014). Use of Fluorescent Techniques to Study the in vitro Movement of Myosins. *Fluorescent Methods Mol. Motors: Exp. Suppl.* **105:** 193–210.

Thomas, L. (1978). *The Lives of a Cell: Notes of a Biology Watcher*. New York, Penguin Books.

Thubrikar, M. (1990). *The Aortic Valve*. Boca Raton, CRC Press, Inc.

Tint, I., D. Jean, P. W. Baas and M. M. Black (2009). Doublecortin Associates with Microtubules Preferentially in Regions of the Axon Displaying Actin-Rich Protrusive Structures. *J. Neurosci.* **29** (35): 10995–11010.

Tognotti, E. (2010). The Eradication of Smallpox, a Success Story for Modern Medicine and Public Health: What Lessons for the Future? *J. Infect. Dev. Ctries* **4** (5): 264–266.

Tomasek, J. J., G. Gabbiani, B. Hinz, C. Chaponnier and R. A. Brown (2002). Myofibroblasts and Mechano-Regulation of Connective Tissue Remodelling. *Nat. Rev. Mol. Cell Biol.* **3** (5): 349–363.

Topp, K. S. and B. S. Boyd (2006). Structure and Biomechanics of Peripheral Nerves: Nerve Responses to Physical Stresses and Implications for Physical Therapist Practice. *Phys. Ther.* **86** (1): 92–109.

Torsvik, V., L. Ovreas and T. F. Thingstad (2002). Prokaryotic Diversity-Magnitude, Dynamics, and Controlling Factors. *Science* **296** (5570): 1064–1066.

Tracqui, P., A. Broisat, J. Toczek, N. Mesnier, J. Ohayon and L. Riou (2011). Mapping Elasticity Moduli of Atherosclerotic Plaque in situ Via Atomic Force Microscopy. *J. Struct. Biol.* **174** (1): 115–123.

Trojer P. and Reinberg D. (2007). Facultative Heterochromatin: Is There a Distinctive Molecular Signature? *Mol. Cell Proteomics* **28** (1): 1–13.

Tseng, Y., J. S. Lee, T. P Kole, I. Jiang and D. Wirtz (2004). Micro-Organization and Visco-Elasticity of the Interphase Nucleus Revealed by Particle Nanotracking. *J. Cell Sci.* **117** (Pt 10): 2159–2167.

Tsilibary, E. C. (2003). Microvascular Basement Membranes in Diabetes Mellitus. *J. Pathol.* **200** (4): 537–546.

Tsukada, M., A. Prokscha, E. Ungewickell and G. Eichele (2005). Doublecortin Association with Actin Filaments Is Regulated by Neurabin Ii. *J. Biol. Chem.* **280** (12): 11361–11368.

Tuszynski, J. A., T. Luchko, S. Portet and J. M. Dixon (2005). Anisotropic Elastic Properties of Microtubules. *Eur. Phys. J. E Soft Matter* **17** (1): 29–35.

Tzvetkova-Chevolleau, T., A. Stephanou, D. Fuard, J. Ohayon, P. Schiavone and P. Tracqui (2008). The Motility of Normal and Cancer Cells in Response to the Combined Influence of the Substrate Rigidity and Anisotropic Microstructure. *Biomaterials* **29** (10): 1541–1551.

Uchida, A., N. H. Alami and A. Brown (2009). Tight Functional Coupling of Kinesin-1a and Dynein Motors in the Bidirectional Transport of Neurofilaments. *Mol. Biol. Cell* **20** (23): 4997–5006.

Uchida, A. and A. Brown (2004). Arrival, Reversal, and Departure of Neurofilaments at the Tips of Growing Axons. *Mol. Biol. Cell* Sep **15** (9): 4215–4225.

Ulrich, T. A., E. M. D. Pardo and S. Kumar (2009). The Mechanical Rigidity of the Extracellular Matrix Regulates the Structure, Motility, and Proliferation of Glioma Cells. *Cancer Res.* **69** (10): 4167–4174.

Vale, R. D., F. Malik and D. Brown (1992). Directional Instability of Microtubule Transport in the Presence of Kinesin and Dynein, Two Opposite Polarity Motor Proteins. *J. Cell Biol.* **119** (6): 1589–1596.

Vallee, R. B. and G. S. Bloom (1991). Mechanisms of Fast and Slow Axonal Transport. *Annu. Rev. Neurosci.* **14**: 59–92.

Van Vliet, K. J., G. Bao and S. Suresh (2003). The Biomechanics Toolbox: Experimental Approaches for Living Cells and Biomolecules. *Acta Mater.* **51** (19): 5881–5905.

Varga, J. and B. Pasche (2008). Antitransforming Growth Factor-Beta Therapy in Fibrosis: Recent Progress and Implications for Systemic Sclerosis. *Curr. Opin. Rheumatol.* **20** (6): 720–728.

Vater, W., W. Fritzsche, A. Schaper, K. J. Bohm, E. Unger and T. M. Jovin (1995). Scanning Force Microscopy of Microtubules and Polymorphic Tubulin Assemblies in Air and in Liquid. *J. Cell Sci.* **108** (Pt 3): 1063–1069.

Vergnes, L., M. Peterfy, M. O. Bergo, S. G. Young and K. Reue (2004). Lamin B1 Is Required for Mouse Development and Nuclear Integrity. *Proc. Natl. Acad. Sci. U. S. A.* **101** (28): 10428–10433.

Wagner, N. and G. Krohne (2007). Lem-Domain Proteins: New Insights into Lamin-Interacting Proteins. *Intl. Rev. Cytol.* **261**: 1–46.

Wagner, O. I., J. Lifshitz, P. A. Janmey, M. Linden, T. K. McIntosh and J. F. Leterrier (2003). Mechanisms of Mitochondria-Neurofilament Interactions. *J. Neurosci.* **23** (27): 9046–9058.

Wahl, M. C. and M. Sundaralingam (1997). Crystal Structures of a-DNA Duplexes. *Biopolymers* **44** (1): 45–63.

Walbeehm, E. T., A. Afoke, T. de Wit, F. Holman, S. E. Hovius and R. A. Brown (2004). Mechanical Functioning of Peripheral Nerves: Linkage with the "Mushrooming" Effect." *Cell Tissue Res.* **316** (1): 115–121.

Walker, G. A., K. S. Masters, D. N. Shah, K. S. Anseth and L. A. Leinwand (2004). Valvular Myofibroblast Activation by Transforming Growth Factor-Beta: Implications for Pathological Extracellular Matrix Remodeling in Heart Valve Disease. *Circ. Res.* **95** (3): 253–260.

Wall, E. J., J. B. Massie, M. K. Kwan, B. L. Rydevik, R. R. Myers and S. R. Garfin (1992). Experimental Stretch Neuropathy. Changes in Nerve Conduction under Tension. *J. Bone Joint Surg. Br.* **74** (1): 126–129.

Walsby, A. E. (1972). Structure and Function of Gas Vacuoles. *Bacteriol. Rev.* **36** (1): 1–32.

Wang, H., B. E. Layton and A. M. Sastry (2003). Nerve Collagens from Diabetic and Nondiabetic Sprague-Dawley and Biobreeding Rats: An Atomic Force Microscopy Study. *Diabetes Metab. Res. Rev.* **19** (4): 288–298.

Wang, H. B., M. Dembo and Y. L. Wang (2000a). Substrate Flexibility Regulates Growth and Apoptosis of Normal but Not Transformed Cells. *Am. J. Physiol. Cell Physiol.* **279** (5): C1345–C1350.

Wang, J., M. Su, J. Fan, A. Seth and C. A. McCulloch (2002). Transcriptional Regulation of a Contractile Gene by Mechanical Forces Applied through Integrins in Osteoblasts. *J. Biol. Chem.* **277** (25): 22889–22895.

Wang, J. H. C. and J. S. Lin (2007). Cell Traction Force and Measurement Methods. *Biomech. Model Mechanobiol.* **6** (6): 361–371.

Wang, L. and A. Brown (2001). Rapid Intermittent Movement of Axonal Neurofilaments Observed by Fluorescence Photobleaching. *Mol. Biol. Cell* **12** (10): 3257–3267.

Wang, L. and A. Brown (2002). Rapid Movement of Microtubules in Axons. *Curr. Biol.* **12** (17): 1496–1501.

Wang, L., C. L. Ho, D. Sun, R. K. Liem and A. Brown (2000b). Rapid Movement of Axonal Neurofilaments Interrupted by Prolonged Pauses. *Nat. Cell Biol.* **2** (3): 137–141.

Wang, M. D., H. Yin, R. Landick, J. Gelles and S. M. Block (1997). Stretching DNA with Optical Tweezers. *Biophys. J.* **72** (3): 1335–1346.

Wang, N. and D. E. Ingber (1994). Control of Cytoskeletal Mechanics by Extracellular Matrix, Cell Shape, and Mechanical Tension. *Biophys. J.* **66** (6): 2181–2189.

Warren, B. A. and J. L. Yong (1997). Calcification of the Aortic Valve: Its Progression and Grading. *Pathology* **29** (4): 360–368.

Waterman-Storer, C. M. and E. Salmon (1999). Positive Feedback Interactions between Microtubule and Actin Dynamics during Cell Motility. *Curr. Opin. Cell Biol.* **11** (1): 61–67.

Watson, J. D. and F. H. C. Crick (1953). A Structure for Deoxyribose Nucleic Acid. *Nature* **171** (April 25): 737–738.

Welte, M. A., S. P. Gross, M. Postner, S. M. Block and E. F. Wieschaus (1998). Developmental Regulation of Vesicle Transport in Drosophila Embryos: Forces and Kinetics. *Cell* **92** (4): 547–557.

Wells, S., and M. Read (2002). *The Journey of Man: A Genetic Odyssey.* Princeton, N.J., Princeton University Press.

Willard, M., M. Wiseman, J. Levine and P. Skene (1979). Axonal Transport of Actin in Rabbit Retinal Ganglion Cells. *J. Cell Biol.* **81** (3): 581–591.

Williamson, T. L., J. R. Marszalek, J. D. Vechio, L. I. Bruijn, M. K. Lee, Z. Xu, et al. (1996). Neurofilaments, Radial Growth of Axons, and Mechanisms of Motor Neuron Disease. *Cold Spring Harb. Symp. Quant. Biol.* **61**: 709–723.

Wipff, P. J., D. B. Rifkin, J. J. Meister and B. Hinz (2007). Myofibroblast Contraction Activates Latent TGF-Beta1 from the Extracellular Matrix. *J. Cell Biol.* **179** (6): 1311–1323.

Wolpert, L. (1991). *The Triumph of the Embryo*. Oxford, England; New York, Oxford University Press.

Wong, J. Y., A. Velasco, P. Rajagopalan and Q. Pham (2003). Directed Movement of Vascular Smooth Muscle Cells on Gradient-Compliant Hydrogels. *Langmuir* **19** (5): 1908–1913.

Wood, J. A., N. M. Shah, C. T. McKee, M. L. Hughbanks, S. J. Liliensiek, P. Russell and C. J. Murphy (2011). The Role of Substratum Compliance of Hydrogels on Vascular Endothelial Cell Behavior. *Biomaterials* **32** (22): 5056–5064.

Worman, H. J. (2005). Components of the Nuclear Envelope and Their Role in Human Disease. *Novartis Found. Symp.* **264**: 35–42; discussion 42–50, 227–230.

Wostyn, P., K. Audenaert and P. P. De Deyn (2008). Alzheimer's Disease-Related Changes in Diseases Characterized by Elevation of Intracranial or Intraocular Pressure. *Clin. Neurol. Neurosurg.* **110** (2): 101–109.

Wynn, T. A. (2007). Common and Unique Mechanisms Regulate Fibrosis in Various Fibroproliferative Diseases. *J. Clin. Invest.* **117** (3): 524–529.

Wyss, K., C. Y. Yip, Z. Mirzaei, X. Jin, J. H. Chen and C. A. Simmons (2012). The Elastic Properties of Valve Interstitial Cells Undergoing Pathological Differentiation. *J. Biomech.* **45** (5): 882–887.

Xia, C. H., E. A. Roberts, L. S. Her, X. Liu, D. S. Williams, D. W. Cleveland and L. S. Goldstein (2003). Abnormal Neurofilament Transport Caused by Targeted Disruption of Neuronal Kinesin Heavy Chain Kif5a. *J. Cell Biol.* **161** (1): 55–66.

Xiao, Y. and G. A. Truskey (1996). Effect of Receptor-Ligand Affinity on the Strength of Endothelial Cell Adhesion. *Biophys. J.* **71** (5): 2869–2884.

Yabe, J. T., A. Pimenta and T. B. Shea (1999). Kinesin-Mediated Transport of Neurofilament Protein Oligomers in Growing Axons. *J. Cell Sci.* **112** (Pt 21): 3799–3814.

Yabuki, S., S. Kikuchi, K. Olmarker and R. R. Myers (1998). Acute Effects of Nucleus Pulposus on Blood Flow and Endoneurial Fluid Pressure in Rat Dorsal Root Ganglia. *Spine (Phila Pa 1976)* **23** (23): 2517–2523.

Yanagida, T., M. Nakase, K. Nishiyama, and F. Oosawa (1984). Direct Observation of Motion of Single F-Actin Filaments in the Presence of Myosin. *Nature* **307**(5946): 58–60.

Yang, X., X. Meng, X. Su, D. C. Mauchley, L. Ao, J. C. Cleveland, Jr. and D. A. Fullerton (2009). Bone Morphogenic Protein 2 Induces Runx2 and Osteopontin Expression in Human Aortic Valve Interstitial Cells: Role of Smad1 and Extracellular Signal-Regulated Kinase 1/2. *J. Thorac. Cardiovasc. Surg.* **138** (4): 1008–1015.

Yap, C. H., N. Saikrishnan, G. Tamilselvan and A. P. Yoganathan (2012a). Experimental Measurement of Dynamic Fluid Shear Stress on the Aortic Surface of the Aortic Valve Leaflet. *Biomech. Model Mechanobiol.* **11** (1–2): 171–182.

Yap, C. H., N. Saikrishnan and A. P. Yoganathan (2012b). Experimental Measurement of Dynamic Fluid Shear Stress on the Ventricular Surface of the Aortic Valve Leaflet. *Biomech. Model Mechanobiol.* **11** (1–2): 231–244.

Ye, Q., I. Callebaut, A. Pezhman, J. C. Courvalin and H. J. Worman (1997). Domain-Specific Interactions of Human Hp1-Type Chromodomain Proteins and Inner Nuclear Membrane Protein Lbr. *J. Biol. Chem.* **272** (23): 14983–14989.

Yeh, W. C., P. C. Li, Y. M. Jeng, H. C. Hsu, P. L. Kuo, M. L. Li, et al. (2002). Elastic Modulus Measurements of Human Liver and Correlation with Pathology. *Ultrasound Med. Biol.* **28** (4): 467–474.

Yeung, T., P. C. Georges, L. A. Flanagan, B. Marg, M. Ortiz, M. Funaki, et al. (2005). Effects of Substrate Stiffness on Cell Morphology, Cytoskeletal Structure, and Adhesion. *Cell Motil Cytoskeleton* **60** (1): 24–34.

Yin, M., J. Woollard, X. F. Wang, V. E. Torres, P. C. Harris, C. J. Ward, et al. (2007). Quantitative Assessment of Hepatic Fibrosis in an Animal Model with Magnetic Resonance Elastography. *Magn. Reson. Med.* **58** (2): 346–353.

Yip, C. Y., M. C. Blaser, Z. Mirzaei, X. Zhong and C. A. Simmons (2011). Inhibition of Pathological Differentiation of Valve Interstitial Cells by C-Type Natriuretic Peptide. *Arterioscler. Thromb. Vasc. Biol.* **31**(8): 1881–1889.

Yip, C. Y., J. H. Chen, R. Zhao and C. A. Simmons (2009). Calcification by Valve Interstitial Cells Is Regulated by the Stiffness of the Extracellular Matrix. *Arterioscler. Thromb. Vasc. Biol.* **29** (6): 936–942.

Yokota, A., M. Doi, H. Ohtsuka and M. Abe (2003). Nerve Conduction and Microanatomy in the Rabbit Sciatic Nerve after Gradual Limb Lengthening-Distraction Neurogenesis. *J. Orthop. Res.* **21** (1): 36–43.

Young, E. W., A. R. Wheeler and C. A. Simmons (2007). Matrix-Dependent Adhesion of Vascular and Valvular Endothelial Cells in Microfluidic Channels. *Lab Chip* **7** (12): 1759–1766.

Young, S. G., L. G. Fong and S. Michaelis (2005). Prelamin a, Zmpste24, Misshapen Cell Nuclei, and Progeria—New Evidence Suggesting That Protein Farnesylation Could Be Important for Disease Pathogenesis. *J. Lipid Res.* **46** (12): 2531–2558.

Young, S. G., M. Meta, S. H. Yang and L. G. Fong (2006). Prelamin a Farnesylation and Progeroid Syndromes. *J. Biol. Chem.* **281** (52): 39741–39745.

Yu, H., L. Ma, Y. Yang and Q. Cui (2007). Mechanochemical Coupling in the Myosin Motor Domain. I. Insights from Equilibrium Active-Site Simulations. *PLoS Comput. Biol.* **3** (2): e21.

Yu, W. and P. W. Baas (1994). Changes in Microtubule Number and Length during Axon Differentiation. *J. Neurosci.* **14** (5 Pt 1): 2818–2829.

Yuan, A., M. V. Rao, Veeranna and R. A. Nixon (2012). Neurofilaments at a Glance. *J. Cell Sci.* **125** (Pt 14): 3257–3263.

Z, S., S. D. and S. AV (2000). Cryoatomic Force Microscopy of Filmanetous Actin. *Biophys. J.* **78** (2): 950–958.

Zaari, N., P. Rajagopalan, S. K. Kim, A. J. Engler and J. Y. Wong (2004). Photopolymerization in Microfluidic Gradient Generators: Microscale Control of Substrate Compliance to Manipulate Cell Response. *Adv. Mater.* **16** (23–24): 2133–2137.

Zastrow, M. S., S. Vlcek and K. L. Wilson (2004). Proteins That Bind a-Type Lamins: Integrating Isolated Clues. *J. Cell Sci.* **117** (Pt 7): 979–987.

Zeiger, A. S. and B. E. Layton (2008). Molecular Modeling of the Axial and Circumferential Elastic Moduli of Tubulin. *Biophys. J.* **95** (8): 3606–3618.

Zhang, Y. (2011). Cargo Transport by Several Motors. *Phys. Rev. E Stat. Nonlin. Soft Matter Phys.* **83** (1 Pt 1): 011909.

Zhao, R., K. L. Sider and C. A. Simmons (2011). Measurement of Layer-Specific Mechanical Properties in Multilayered Biomaterials by Micropipette Aspiration. *Acta Biomater.* **7** (3): 1220–1227.

Zhao, X. H., C. Laschinger, P. Arora, K. Szaszi, A. Kapus and C. A. McCulloch (2007a). Force Activates Smooth Muscle Alpha-Actin Promoter Activity through the Rho Signaling Pathway. *J. Cell Sci.* **120** (Pt 10): 1801–1809.

Zhao, Y., L. N. An and J. Y. Fang (2007b). Buckling of Lipid Tubules in Shrinking Liquid Droplets. *Nano Lett.* **7** (5): 1360–1363.

Zheng, J., R. E. Buxbaum and S. R. Heidemann (1994). Measurements of Growth Cone Adhesion to Culture Surfaces by Micromanipulation. *J. Cell Biol.* **127** (6 Pt 2): 2049–2060.

Zheng, J., P. Lamoureux, V. Santiago, T. Dennerll, R. E. Buxbaum and S. R. Heidemann (1991). Tensile Regulation of Axonal Elongation and Initiation. *J. Neurosci.* **11** (4): 1117–1125.

Zhu, Q., S. Couillard-Despres and J. P. Julien (1997). Delayed Maturation of Regenerating Myelinated Axons in Mice Lacking Neurofilaments. *Exp. Neurol.* **148** (1): 299–316.

Index